AF322579

Introduction to Marine Plankton

Dr. Abhijit Mitra, a Faculty Member of the Department of Marine Science, University of Calcutta has been active in the sphere of Marine Biology since 1985. He secured the first position in the first class in M. Sc examination according to the order of merit and obtained his Ph.D degree as NET qualified scholar from the University of Calcutta. He served as a Senior Research Fellow of Calcutta Port Trust to monitor the impact of pollution on mangrove biodiversity. He also served as a Scientist in the Ministry of Environment and Forests project, Government of India to evaluate the fish juvenile loss due to wild harvest of prawn seeds for meeting the demand of shrimp culture farms in coastal West Bengal. In 1997, he joined the World Wide Fund for Nature – India (Eastern Region) as a Senior Programme Officer. He took the charge of co-ordinator in the World Bank Funded project on the Leopard depredation problem in the North Bengal during the tenure of his service at WWF – India (Eastern Region). He provided consultancy services to several environmental impact assessment projects like Bakreshwar Thermal Power Project, Mumbai Trans-Harbour Link (MTHL) project, feasibility study for setting up shrimp culture farms in the North East coast of the Bay of Bengal, fish landing stations in coastal West Bengal etc. In 2002, he was selected as a pisciculture specialist for formulating a project on livelihood improvement in Sundarbans under Asian Development Bank (ADB). Dr. Abhijit Mitra is a member of many scientific committees in India and abroad and was invited to chair a session on Mangrove Macrobenthic conference at Mombasa, Kenya (during 7th to 11th September, 2000). He has to his credit about 125 papers in various National and International journals. He is also the author of 10 books in the sphere of Environmental science.

Ms. Kakoli Banerjee is well known in the sphere of Environmental Science and is presently associated as a Scientist in a Coastal Zone Management Project of Government of West Bengal. She has published about 25 research papers in various National and International journals. She has also contributed some important chapters in some reputed books like Ecology and Ethology of Aquatic Biota, Ecology of Polluted Waters, etc. She received the "Best Basic Research Award" for her outstanding findings in the sphere of phytoplankton taxonomy. She is associated with the ecological assessment component for setting up fish landing stations in coastal West Bengal. She is a member of various professional bodies, societies and committees like Society of Indian Ocean Studies (SIOS), Creative Research Group (CRG), Indian Association for the Cultivation of Science (IACS), S.D. Marine Biological Research Institute (SDMBRI) and Central Calcutta Science and Culture Organisation for the Youth (CCSCOY).

Dr. Avijit Gangopadhyay, is presently serving as the Associate Professor of Department of Physics and School for Marine Science and Technology (SMAST), UMass Dartmouth. He is also the Faculty Member of Intercampus Graduate School of Marine Science and Technology, University of Massachusetts, USA. He is the Honorary Research Associate of Harvard University, Cambridge MA. He completed his Ph. D (Ocean Engineering) in 1990 from University of Rhode Island, Kingston and has a wide range of research experience in several internationally reputed laboratories like Applied Physics Laboratory of Johns Hopkins University, Ocean Engineering Division of Engineers India Limited, New Delhi, Harvard University etc. He has 30 publications in various National and International journals and served as investigators in several International projects. He is the member of the American Geophysical Union, American Meteorological Society and Sigma XI. Dr. Gangopadhyay is the reviewer for many reputed journals.

Introduction to Marine Plankton

Abhijit Mitra

Kakoli Banerjee

Avijit Gangopadhyay

2020

Daya Publishing House®

A Division of

Astral International Pvt. Ltd.

New Delhi – 110 002

Published by   : **Daya Publishing House®**
A Division of
Astral International Pvt. Ltd.
– ISO 9001:2015 Certified Company –
4736/23, Ansari Road, Darya Ganj
New Delhi-110 002
Ph. 011-43549197, 23278134
E-mail: info@astralint.com
Website: www.astralint.com

Contents

Chapter 1
MARINE & ESTUARINE PLANKTON : AN OVERVIEW

Contents

1.1 Habitat of marine and estuarine plankton

The sea water is the dwelling place for a large variety of organisms. However it is like a multi-storied building, where the inhabitants of the first floor are not exactly similar to the inhabitants of the 29th or 31st floor. The entire oceanic water (pelagic system) is similarly divided into several tiers like epipelagic zone (upto 200 m), mesopelagic zone (200 m to 1000 m), bathypelagic zone (1000 m to 4000 m) and abyssopelagic zone (4000 m to 6000 m). The planktonic species of one zone has least liking for the other zone (a case of niche preference). Most of the planktonic organisms are restricted in the neritic zone due to abundance of nutrients, light and favourable physico-chemical variables (like water temperature, dissolved oxygen, salinity, pH etc.).

The word plankton has come from the Greek word *"planktos"*, which means a passively drifting or wandering form. Depending upon whether a planktonic organism is a plant or an animal, a distinction is made between **phytoplankton** and **zooplankton**. Although many planktonic species are of microscopic dimensions, the term is not synonymous with small size as some of the zooplankton include jellyfish of several meters in diameter. It is not necessary that all phytoplankton are completely passive, many of them are capable of swimming too.

Phytoplankton are free floating tiny floral components that are widely distributed in the marine and estuarine environments. Like land plants, these tiny producers require sunlight, nutrients or fertilizers, carbon dioxide gas and water for their growth and survival. The cells of these organisms contain the pigment **chlorophyll** that traps the solar energy for use in **photosynthesis**. The photosynthetic process uses the solar radiation to convert carbon dioxide and water into sugars or high energy organic compounds from which the cell forms new materials. The synthesis of organic material by photosynthesis is termed **primary production**. Since phytoplankton are the dominant producers in the ocean, their role in the marine food chain is of paramount importance. Approximately, 4000 species of marine phytoplankton have been described and new species are continually being added to this total (Lalli & Parsons, 1997). Phytoplankton exhibit remarkable adaptation to remain in floating condition in the saline soup. In fact, all marine phytoplankton tend to stay in the photic zone to utilize the solar radiation for performing the process of photosynthesis. In order to retard the process of sinking, various mechanisms are adopted by this group of organisms. These include their small size and general morphology, as the ratio of cell surface area to volume determines frictional drag in the water. Colony or chain formation also increases surface area and slows the rate of sinking. Most species carry out ionic regulation, in which the internal concentration of ions is reduced relative to their concentration in seawater. Diatoms also produce and store oil and this metabolic by-product further reduces cell density. In nature, turbulence of surface waters is also an important physical factor in maintaining phytoplankton near the surface where they receive abundant sunlight.

Phytoplankton play an important role in maintaining the global carbon cycle. The elemental composition of phytoplankton is C:N:P = 106:16:1, which is commonly referred to as the "Redfield Ratio". This ratio clearly explains that about 100 units of carbon is delivered to the deep sea for every 16 units of nitrogen and 1 unit of phosphorus. As such, the biological pump delivers carbon from the atmosphere to the deep sea compartment, where it is concentrated and sequestered for centuries. The green house effect is thus controlled by these tiny floral components of the sea.

The zooplankton occupy the tier next to phytoplankton in the marine food web. They either graze on phytoplankton or feed on other members of zooplankton. Foraminifera and radiolarians are members of zooplankton community who possess only one cell. Copepods and euphausiids are widely distributed zooplankton in the marine and estuarine environments. Zooplankton exhibit the phenomenon of **vertical migration** in which some species migrate towards the sea surface during the night time and return back to their original depth during day. The amount of daily migration varies between 10 m and 500 m and this change of zone is keenly related to their nutritional requirement or maintenance of their light level.

The zooplankton are mainly filter feeding in nature and are the prey of many shore-dwelling species. The zooplankton also exhibit various adaptations to keep them in floating condition. In many protozoa, the cilia act as blade of the rotor of a helicopter which enable the ciliates to move up and down the water. The scyphomedusae floats and moves in the direction of current by the rhythmic contraction and relaxation of their umbrella. Although these are large and heavy, but 99% of their mass is due to the presence of a jelly-like substance, due to which their density is reduced. *Vellela* sp. has chambers filled with air that keeps it afloat. The zoea of many arthropods have well developed uropods and these help them to remain in floating condition or buoy up by swimming. The molluscan larvae are important component of zooplankton community. The *Ianthina* sp. remains afloat by creating froth of bubbles, which contract the surface film of water and the animal hangs on froth. The zooplankton usually produce three to five generations a year in warm water, where there is plenty of food supply and optimum temperature. At higher latitudes, where the phytoplankton growth is restricted within a brief period of time, the zooplankton produce only a single generation in a year.

1.2. Importance of phytoplankton and zooplankton

The marine and estuarine food webs are spun with the energy of plankton. The fascinating creatures like dolphins, whales or other marine mammals, the school of sharks or the benthic oysters in the intertidal zone are all dependent on the plankton either directly or indirectly for their food and nutrition. Phytoplankton form the foundation stone of world fishery. It has been estimated that the total annual primary production of the world seas is around 20 x 10⁹ tonnes of carbon which has the capacity to yield 240 million tonnes of fish although our present harvest is only about 60 million tonnes.

The distribution and abundance of the commercially important fish and shellfish and their larvae are dependent on some species of the phytoplankton besides serving as their main food source. Among the species of diatoms, *Fragilaria oceanica* and *Hemidiscus hardmannianus* have been recorded to indicate the abundance of the clupeid fish, Hilsa in the Hooghly estuary and oil sardine, *Sardinella longiceps* in the west coast of India. The **'white water phenomenon'** due to abundance of the coccolithophores has been a good omen for the herring fisheries in the British waters. Similarly the colonial diatom, *Fragilaria antarctica* is known to indicate the abundance of the antarctic krill, *Euphausia superba*.

Although elaborate studies have been made on marine and estuarine phytoplankton in many countries of the world, knowledge of this group of marine flora is very scanty in India. Hence, large scale investigations on the phytoplankton especially on their dynamics in Indian seas are imperative to locate new fishing grounds and to understand their role in sustaining the blue revolution.

Phytoplankton are the sources of fossil fuels. It has been reported that some species of phytoplankton store their extra food as oil rather than starch to get an advantage in terms of buoyancy. When oil-storing forms of phytoplankton die and descend to the bottom of the sea and undergo microbial degradation, they are often covered by huge sediment load and subjected to enormous pressure. Such conditions are supposed to change the deposited phytoplankton into fossil fuel after hundreds and millions of years. Thus human world is benefitted from these tiny organisms even after their death.

The diatoms are commercially important as **'diatomaceous earth'** (diatomite or kieselghur), a congregation of dead silicon-rich diatom frustules in the sea beds like protozoan oozes. This material is employed in the filtration of fruit juice, syrup and varnish; in the boilers, electric ovens and refrigerators as insulators; as polishing agent for precious metals; in the separation of paraffin wax from petroleum and in the manufacture of concrete. Because of the presence of the hydrocarbon compound **'diatomin'**, the diatoms serve as the indicators of rich petroleum grounds. Present day scientists are of the opinion that these tiny free-floating producers are the chief precursors of the petroleum rich fields of Venezuela at Los Angeles.

The dinoflagellates also at times play an important role in the fishery sector as by virtue of their power of luminescence they can emit 'cold living light' which though lasts for about one-tenth of a second, helps in identifying certain fish shoals during night.

Phytoplankton can also be used to reflect the water quality in terms of certain pollutants other than sewage (that contribute considerable amount of nitrate and phosphate in the ambient water). It has been reported that an increase in cell volume occurs in certain species (like *Dunaliella tertiolecta* and *Phaeodactylum tricornutum*) due to copper pollution (Stauber and Florence, 1987).

Phytoplankton present in the marine and estuarine environments use carbon dioxide for photosynthesis and hence play an important role in maintaining the carbon dioxide budget of the atmosphere. The larger the world's phytoplankton population, the more carbon dioxide gets pulled from the atmosphere. This lowers the average temperature of the atmosphere due to lower volumes of this greenhouse gas. Scientists have found that a given population of phytoplankton can double its numbers in the order of once per day. In other words, phytoplankton respond very rapidly to changes in their environment. Large populations of this organism, sustained over long periods of time, could significantly lower atmospheric carbon dioxide levels and in turn, lower average temperatures.

Diatoms are important components contributing to the stability of intertidal mudflats. They excrete polysaccharides, which trap sediment grains and stabilize the sediment **(http://www.nioo.know.hl/cemo/ecoflat/work.htm)**.

Mass culture of phytoplankton may be undertaken in closed systems of the space capsules. The nitrogenous wastes of the travellers could be diluted and circulated as algal nutrients and the respired air as a source of carbon dioxide. In exchange, the travellers could utilize the oxygen released by the photosynthesis of algae and benefit from the food value of cultured phytoplankton.

There are major contributions of zooplankton to the marine and estuarine environments. Among the various levels of production in the sea, the secondary production contributed by zooplankton is an important linkage between the primary and tertiary productions. The zooplankton mainly consume the primary producers and form the major food source for tertiary producers. An important food for the antarctic baleen whales are the krills which are important zooplankton in the antarctic region.

In certain parts of the world, the zooplankton are directly consumed as food, e.g., certain species of *Mysids* are consumed as food in West Indies. A deep water copepod (*Euchaeta norvegica*) is a delicious food item in West Indies. In many corners of the globe, zooplankton are treated as source of bioactive substances. Certain species of zooplankton are used as bioindicators of water quality. The zooplankton contribute substantial biogenic material to ooze formation, which has wide application in instrument related industry where the ooze is used as thermal insulators, chromatographic column filters etc.

Suggested references

1. Stauber, J. L. and T. M. Florence (1987). Mechanism of toxicity of ionic copper and copper complexes to algae. Marine Biology, 94 : 511-519.

2. Lalli, G. M. and T. R. Parsons, (1997). Energy flow and nutrient cycling, p. 112-146. In: Biological Oceanography: an introduction, 2nd ed. Open University.

3. WWF, Marine Update 50. (August, 2001). A Report on Eutrophication and European Marine sites.

4. Lazaroff, N. and Vishniac, W. (1962). The participation of filament anastomosis in the development cycle of *Nostoc muscorum*, a blue-green alga. J. Gen. Microbiol., 28: 203-210.

Internet references

- *http://www.bigelow.org/sci_overview*
- *http://www.nioo.know.hl/cemo/ecoflat/work.htm)*
- *http://oceanlink.island.net/ask/pollution.html)*
- *http://www4.fimr.fi/algaline/sheets/algasyst*
- *http://www.gek.org/phytoa.htm*
- *http://www.mos.org/oceans/life/surface.html*

Chapter 2
Marine & Estuarine Plankton : Classification & Types

Contents

2.1   Classification of marine and estuarine phytoplankton

Very minute free-floating plants that drift in the aquatic phase are usually referred to as phytoplankton. They are found in either shallow or deep aquatic system where there is availability of light. They utilize the light energy for producing organic compounds through the process of photosynthesis. Phytoplankton form the primary biological component of any aquatic system, sweet or saline, from which energy is transferred to higher trophic levels occupied by various species of vertebrates (like fishes) and invertebrates. In comparison to terrestrial plants and submerged macrophytes, the populations of this tiny drifing plant community turn out on a much shorter scale, blooms being of only few weeks duration and species replacement occurs only within a month or less. The fertility of the marine and estuarine systems is reflected through the bioproductivity of the phytoplankton. Marine phytoplankton, as primary producers and zooplankton, as secondary producers are the key players in spinning the marine food webs in which large fishes and marine mammals also exist.

Nearly all marine plants, whether unicellular or multicellular, even those attached to substrata (sessile) or free floating, pass some part of their life cycle in floating condition as phytoplankton. However, those organisms which always remain planktonic through out the life cycle are 1) diatoms, 2) dinoflagellates, 3) coccolithophores, 4) selective species of blue-green algae and 5) some species of green algae.

Diatoms

These floating plants are all microscopic in size and are characterised by the presence of shell or frustule. The shell or frustule is composed of translucent silica. The cell wall of diatom has two parts resembling a pillbox bottom and lid. The lid is called the **epitheca** and the bottom is known as **hypotheca**. These shells have great importance from the geological point of view and constitute the diatomaceous crust. The diatoms exhibit remarkable varieties and forms and many species possess beautifully sculptured shells (Fig. 2.1.1).

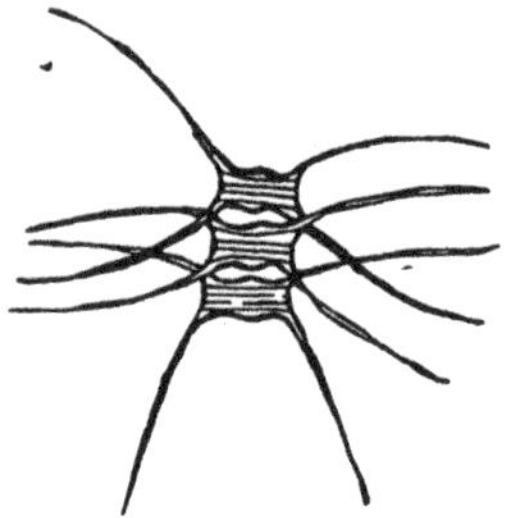

Bacteriastrum hyalinum   *Chaetoceros lorenzianus*

Fig. 2.1.1. Some common diatoms of marine and estuarine environments

Depending on the nature of valves and pattern of ornamentation in the valve surface, the diatoms are grouped into **centric** and **pennate** diatoms. The major differences between these two groups are given in table 2.1.1.

TABLE 2.1.1 : Differences between centric and pennate diatoms

Point	Centric diatom	Pennate diatom
Cell shape	Discoid, solenoid or cylindircal.	Elongated and fusiform oval, sigmoid or roughly circular.
Ornamentation	Radial in nature i.e., the arrangement of markings is radiating from the centre.	Bilateral in nature i.e., the arrangement of the markings is on either side of the apical (main) axis.

Dinoflagellates

These are important producers of the marine and estuarine environments and rank second according to order of importance in the economy of the sea. Typically, these are unicellular - some are naked while others are armoured with plates of cellulose. The dinoflagellates possess two flagella for locomotion. Several of them are luminescent and produce light (Fig. 2.1.2).

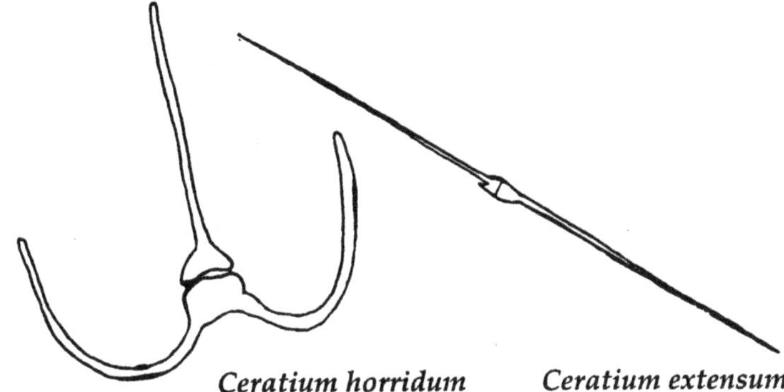

Fig. 2.1.2. Some common dinoflagellates of marine and estuarine environments

Coccolithophores

These are among the smallest category of phytoplankton having a size range between 5 to 20 microns. Some coccolithophores have flagella while others are devoid of them. Their soft bodies are shielded by tiny, calcified circular plates or shields of various design (Fig. 2.1.3). These are normally found in the open sea, but their profuse occurrence has been recorded in coastal waters. They form important diet components of filter feeding animals.

Fig. 2.1.3. Representatives of coccolithophores

Blue-green algae

These include both unicellular or multicellular organisms. The blue colour in them is due to the presence of a pigment known as phycocyanin. Of the various organisms belonging to this category, the most important is *Trichodesmium erythraeum* because in certain seasons of the year its biomass increases greatly resulting in the formation of clumps. Some common phytoplankton belonging to this category are shown in figure 2.1.4.

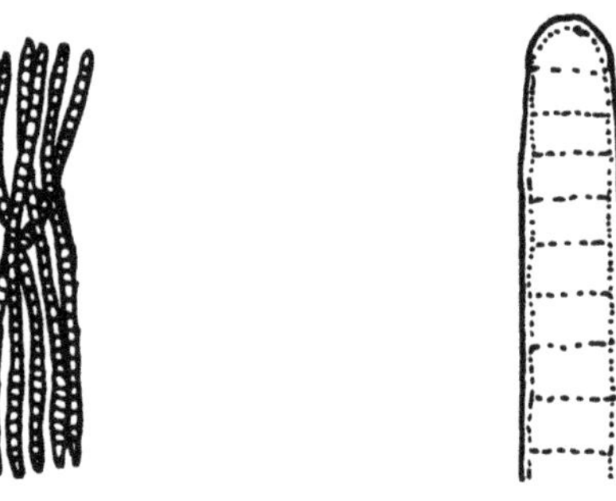

Trichodesmium erythraea   *Oscillatoria sp.*

Fig. 2.1.4. Representatives of planktonic blue-green algae

Green algae

Microscopic green algae are important component of the planktonic community that largely occur in coastal waters. The green colour in them is due to the presence of chloroplasts. They are widely distributed in the warmer (tropical) seas and only few species are found in the Arctic and Antarctic oceans. Some common species of microscopic green algae that are planktonic in nature are shown in figure 2.1.5.

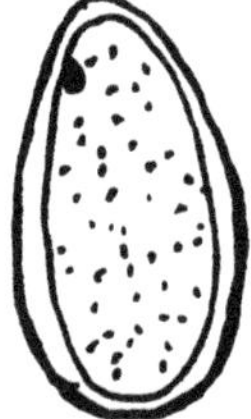

Chlorella salina   *Chlorella marina*

Fig. 2.1.5. Some common planktonic green algae

The phytoplankton community consists of a variety of groups namely diatoms, dinoflagellates, blue-green algae, silico-flagellates, coccolithophores etc. whose members range in terms of size from 0.001 to 0.2 mm. There are various approaches to classify the vast spectrum of species existing in the phytoplankton community. Some workers have classified this community on the basis of size, while some have preferred to divide the species on the basis of cell characteristics.

a) On the basis of size, the phytoplankton may be grouped under five categories (Table 2.1.2).

TABLE 2.1.2 : Classification of phytoplankton on the basis of size

Plankton category	Maximum dimension
Ultraplankton	<2 µm
Nannoplankton	2-20 µm
Microplankton	20-200 µm
Macroplankton	200-2000 µm
Megaplankton	>2000 µm

b) Phytoplankton may also be classified on the basis of the cell characteristics (Table 2.1.3).

TABLE 2.1.3 : Classification of phytoplankton on the basis of cell characteristics

Class	Common name	Area(s) of predominance	Common genera
Cyanophyceae (cyanobacteria)	Blue-green algae	Tropical	*Oscillatoria, Synechococcus*
Rhodophyceae	Red algae	Cold temperate	*Rhodella*
Cryptophyceae	Cryptomonads	Coastal	*Cryptomonas*
Chrysophyceae	Chrysomonads	Coastal	*Aureococcus*
	Silicoflagellates	Cold waters	*Dictyocha*
Bacillariophyceae (Diatomophyceae)	Diatoms	All waters, especially coastal waters	*Coscinodiscus, Chaetoceros, Rhizosolenia*
Raphidophyceae	Chloromonads	Brackish	*Heterosigma*
Xanthophyceae	Yellow-green algae	Brackish	Very rare
Eustigmatophyceae	Yellow-green algae	Estuarine	Very rare
Prymnesiophyceae	Coccolithophorids	Oceanic	*Emiliania*
	Prymnesiomonads	Coastal	Isochrysis *Prymnesium*
Euglenophyceae	Euglenoids	Coastal	*Eutreptiella*
Prasinophyceae	Prasinomonads	All waters	*Tetraselmis Micromonas*
Chlorophyceae	Green algae	Coastal	Rare
Pyrrophyceae (Dinophyceae)	Dinoflagellates	All waters, especially warm	*Ceratium Gonyaulax Protoperidinium*

Source: Lalli and Parsons, 1997

2.2.   Classification of marine and estuarine zooplankton

Zooplankton are found in almost all the layers of the photic zone of the ocean. They are potentially limited by two factors in the coastal and estuarine sectors. Firstly by turbidity, which can limit phytoplankton production and thus restrict the ration supply for the zooplankton community, and secondly by currents which, particularly in small estuaries are dominated by high river flow, that usually carry the members of zooplankton out to the sea. The zooplankton biomass can increase the fishery productivity because they chiefly consume the primary producers (phytoplankton) and form the major food source for members of higher trophic levels in which several species of osteichthyes and chondrichthyes are existing.

Omori and Ikeda (1984) classified the zooplankton according to their habitat, depth distribution, size and duration of planktonic life (Tables 2.2.1, 2.2.2, 2.2.3 and 2.2.4).

TABLE 2.2.1 : Classification of zooplankton on the basis of habitat

Type	Description
Oceanic plankton	These are marine zooplankton that inhabit beyond the continental shelf.
Neritic plankton	These zooplankton inhabit waters overlying continental shelves. These waters are often very productive as they receive the run-off from the adjacent landmasses that triggers the phytoplankton growth in these regions.
Brackish water plankton	These zooplankton inhabit estuarine regions, where there is a continuous mixing of fresh water and sea water. The zooplankton species of this category have wide range of tolerance to different dilution factors. Such zooplankton are very common in the shrimp culture farms and form important diet of the prawns.

TABLE 2.2.2 :Classification of zooplankton on the basis of depth distribution

Type	Description
Neuston	The zooplankton of this category are restricted at the top few millimeters (usually 10 mm) of the surface micro layer.
Pleuston	These zooplankton are widely distributed at the surface of the sea (with parts of the body some time projecting above the water).
Epipelagic	These zooplankton are distributed between 0 to 300 m water column e.g., siphonophores, arrow worms etc.
Mesopelagic	The zooplankton of this category are restricted within the depth 300 m to 1000 m, e.g., euphausiids, chaetognath etc.
Bathypelagic	These zooplankton are restricted within the depth 1000 m and 3000 m e.g., foraminifera, euphausiids etc.
Abyssopelagic	The waters overlying the vast abyssal plains of the ocean are inhabited by a variety of zooplankton species which are often referred to as abyssopelagic zooplankton. These zooplankton are thus restricted between 3000 m to 4000 m.

Source: Santhanam and Srinivasan, 1998

TABLE 2.2.3 : Classification of zooplankton on the basis of size

Type	Size range
Nannozooplankton	$< 20\ \mu m$
Microzooplankton	$20 - 200\ \mu m$
Mesozooplankton	$200\ \mu m - 2\ mm$
Macrozooplankton	$2 - 20\ mm$
Megazooplankton	$> 20\ mm$

Source: Santhanam and Srinivasan, 1998.

TABLE 2.2.4 : Classification of zooplankton on the basis of duration of planktonic life

Type	Description
Holoplankton	This group includes organisms which are planktonic through out their life cycle e.g., tintinnids, cladocerans, copepods, chaetognaths etc.
Meroplankton	This group encompasses those organisms, which remain planktonic only for a portion of their life cycle e.g., larvae of benthic invertebrates and fish larvae (ichthyoplankton)

Source: Santhanam and Srinivasan, 1998.

Suggested references

1. Banerjee, Kakoli; Mitra, Abhijit; Bhattacharyya, D. P. & Choudhury, Amalesh (2000). A preliminary study of phytoplankton diversity and water quality around Haldia port-cum-industrial complex. Proceedings of the National Seminar on "Protection of the Environment – An urgent need". IPHE, Kolkata.

2. Lalli C. M and Parsons T. R. (1997). In: Biological Oceanography – An Introduction. 2nd Edition. The Open University Set Book.

3. Santhanam, R and Srinivasan, A (1998). In: A manual of marine zooplankton. Oxford and IBH publishing Co. Pvt. Ltd., 1-4.

Internet references

- *http://www.marbot.gu.se/SSS/SSShome.htm*
- *http://www.marbot.gu.se/SSS/dinoflagellates/dino-frame.htm*
- *http://www.marbot.gu.se/SSS/diatoms/diatom_frame.htm*

Chapter 3
SAMPLING AND STORING OF PLANKTON

Contents

3.1. Sampling and preservation

Plankton sampling is done from sea water by using special types of nets through which large volumes of water are filtered to concentrate the planktonic organisms. The plankton net consists of a cone-shaped gauze bag equipped with a metal ring at the wider end and closed at the narrow end by a detachable plankton collecting bottle of a known volume. The gauzes used in nets are made of different materials such as bolting silk, polyester, nylon etc. High quality phytoplankton is retained if the mesh size of the net is less than 20 µm. Depending on the fineness of the filtering portions, a number of monofilament nylon materials are available (Table 3.1.1). The number of the net material and mesh apertures per unit area of net increase with an increase in fineness of the net.

TABLE 3.1.1 : Observed values of mesh widths of various monofilament nylon net materials

Net No.	Average mesh width (in μm)
0	508
3	306
4	288
8	184
12	130
13	115
14	101
15	93
16	88
17	78
18	72
19	70
20	65
21	62
25	55
30	41

Usually the surface samples for plankton analysis are obtained with a clean plastic bucket of measured volume and water is ultimately passed through the net. In shallow brackish water, shrimp culture farms and in lagoons, water samples are obtained using a weighed flexible or rigid plastic tube. To the other end of the rigid tube, a PVC ball valve is mounted to exclude the air. The tube is sent down vertically up to a measured depth and then closed at the top to trap a core of water. It is finally drawn up and by opening the valve or the clamp air is let in. The sample is ultimately collected in a container and passed through the net.

Today's oceanographer's sample zooplankton with multiple net systems mounted on a single frame. These nets can be closed or opened as per the command given from the research vessel. The frame also carries electronic sensors that relay data on salinity, water temperature, water flow, light level, net depth and the angle of the tow to the computer of the research vessel.

For subsurface samples from deeper waters several discrete water samplers such as Van Dorn samplers, Niskin bottles (5 litre, 8 litre or 20 litre) or non-toxic PVC samplers or go-flow samplers are used. Description of some common plankton samplers are given here. However the water collected through all these samplers are finally passed through the plankton net and the concentrated sample is collected within the bottle attached to the narrow end of the net.

A] Meyer's water sampler

It is the simplest instrument for collecting the water samples from any desired depth of shallow systems like the nearshore waters, estuaries and mangroves, where there is no appreciable hydrostatic pressure.

The sampler consists of an ordinary glass or perspex bottle of about 2 litre capacity and is enclosed with a metal band. It is made heavier by providing a lead weight and there are two strong nylon graduated ropes, one tied to the neck of the bottle and the other to the cork (Fig. 3.1.1).

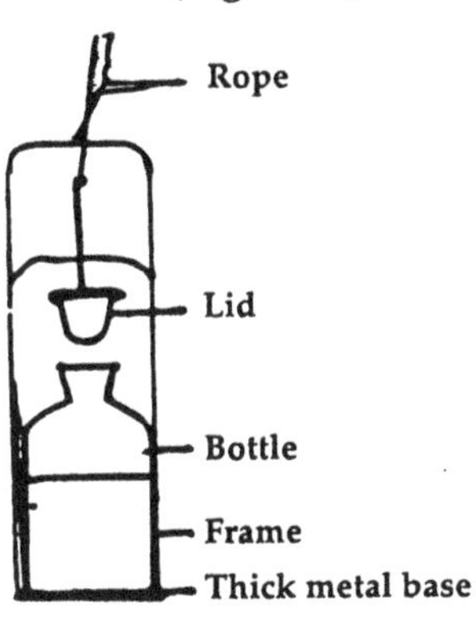

Fig. 3.1.1

During operation of the sampler, the closed bottle is allowed to sink to the desired depth by using the neck rope, where the stopper is jerked open by a strong pull of the cork rope. At this stage, the water flows into the bottle. When the bottle becomes full, which is known by the disappearance of air bubble, the cork rope is released to close the cork after filling sample water inside the bottle. Finally, the bottle containing the water sample is drawn out of the water column by using the neck rope. This type of sampler can effectively function at a depth of up to 20 m.

B] Friedinger's water sampler

It is used to collect the water samples together with the planktonic organisms from a desired depth. The sampler has a cylindrical portion made of plexiglass or perspex and is provided with two hinged covers (Fig. 3.1.2).

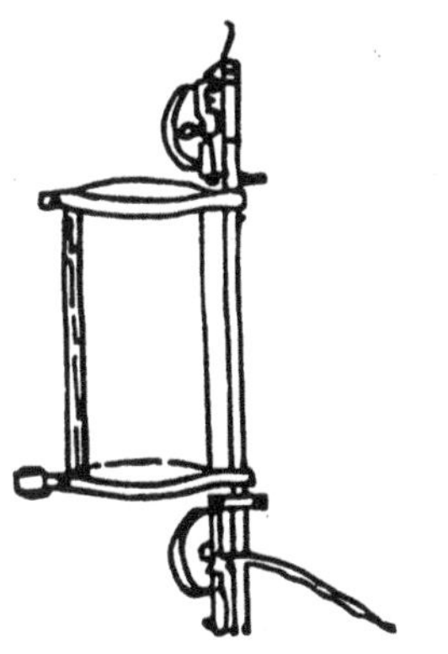

Fig. 3.1.2

During operation, the sampler is sent down in an open state to the desired depth and closed by dropping down the weight messenger. The weight messenger falls down inside a sliding rail and closes the covers and makes the bottle water-tight. By this way, the water together with the planktonic organisms of the desired column is trapped inside the sampler.

C] Nansen reversing water sampler

It is also used widely for collecting water samples for plankton enumeration from subsurface levels. The reversing bottle is made of brass and plated inside with tin or silver or coated with a special lacquer. It is fitted with a drain cock and an air-vent to facilitate draining the trapped water sample. The reversing water samplers are available in different capacities (750 - 1250 ml). Each sampler is also provided with two plug valves, one on each end of the metal cylinder and are operated synchronously by means of a connecting rod fastened to the clamp which secures the sampler to the wire rope (Fig. 3.1.3).

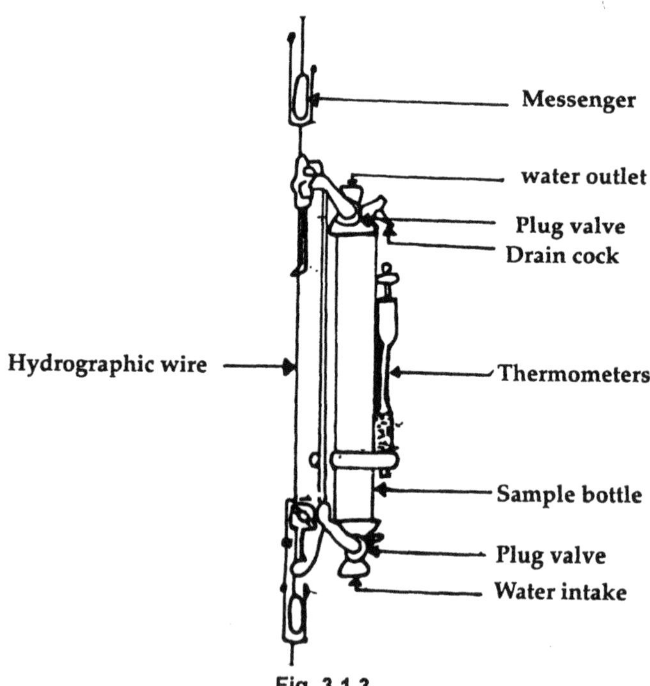

Fig. 3.1.3

When the sampler is lowered, the clamp at the lower end and the plug valves remain in open condition, so that water can pass through the sampler. The sampler is held in this position by the release mechanism which passes around the wire rope. However when the messenger is dropped down the rope, it strikes hard and the sampler immediately falls over and turns through 180^0, shutting the valves which are then closed by a locking device. Finally the water sample of the desired depth is pulled up onto the vessel in a closed condition and passed through the plankton net to concentrate the tiny organisms in the collecting bottle attached at the end of the conical net. The concentrated plankton samples are then preserved for taxonomic identification.

The process of fixation and preservation of plankton are done by various chemicals as discussed here.

FORMALIN

It is a widely used fixative and preservative applied for preservation of a variety of organisms including plankton. The commercial formalin is obtained as a 40% (saturation limit) formaldehyde dissolved in water. Some menthol content is also used as a fixative, though it is a poisonous substance.

Formalin generally reacts with metal and hence is always stored in inert glass or plastic containers. In ordinary use, formalin is stored in amber coloured bottles and kept in cool temperature. If kept in light coloured bottles and exposed to sunlight, a poisonous white precipitate is developed due to the formation of paraformaldehyde. The commercial formalin also contains dissolved impurities such as iron and formic acid which disintegrate the shells of some planktonic organisms. Impure formalin shows the presence of iron which comes out as a brown flocculent precipitate in which the plankton get entangled. Such precipitation of iron can be prevented by adding Rochelle salt (sodium potassium tartarate) at the rate of 10 gm salt to 1 litre of 40% commercial formalin. The acidity of commercial formalin is reduced by adding calculated amount of $CaCO_3$.

LUGOL'S SOLUTION

This preservative is prepared by adding 100 gm of KI in 1 litre of distilled water, 50 gm of iodine (crystalline) and 100 ml of glacial acetic acid. This type of preservative is used for all phytoplankton except coccolithophorids as the acid may dissolve the coccoliths. The preservation is done by adding 0.4 to 0.8 ml of fixative in 200 ml sample.

OSMIC ACID

200 mg of osmium tetroxide added to 10 ml of distilled water leads to the formation of this preservative. This preservative is added at the rate of 3-6 drops per 100 ml phytoplankton sample.

GLUTERALDEHYDE

The preparation of this solution is done by mixing 8 gm of gluteraldehyde in 100 ml of distilled water. The application of this preservative is done in the ratio of 1:1.

Phytoplankton samples especially diatoms are stored in polythene bottles. If they are stored in glass bottles or other metal containers there is a chance of damage of their silica shells in due course of time.

The process of staining planktonic species is very specific. Vital and mortal stains such as neutral red and Evans blue are commonly used to stain plankton. Fluorochromes are used to enhance fluorescence quantum yield, particularly in bacteria. The staining technology developed by various workers are discussed here in brief:

- DeNoyelles (1968) developed a stained – organism filter technique. Samples are preserved with formalin and acetic acid in a 5:2 ratio. To 93 ml of sample 7 ml of preservative is added. Aniline blue and eosin Y stains are prepared separately by the addition of 0.7 gm of dry stain to 50 ml water. One drop of each stain is added to 50 ml sample. The sample is filtered using a syringe that holds a 13 mm diameter MF-Millipore HA filter. The filter is taken out and placed on a glass slide. It is cleared by rinsing with a few drops of diluted Einschlu mittel W 15. The filter is then mounted in the same. Organisms after this process of staining take either red or blue colour.

- Reynolds *et al* (1978) recommended the use of 0.02 ml of a 0.1 % solution (w/v) of neutral red to 1 ml aliquot and 0.05 ml of a 1 % solution (w/v) of Evans blue to 1 ml aliquot. Samples are examined after 30 min. Evans blue, an acidic dye is effective as a mortal stain. Life stains are stained brilliant blue very rapidly which lasts for longer period in the cell. Of the 8 estuarine cultures tested, this method was suitable in 6 species (Subba Rao, 2000). In most cases, fixation with 3 % buffered formalin did not effect the cells.

- According to Bentley-Mowat (1982), hydrolysis by fluorescein diacetate (FDA) enhances the green fluorescence and is widely used on marine phytoplankton. A stock solution of the fluorochrome (0.5% in acetone) is prepared and used as 0.01% solution in freshly filtered seawater at 0°C. Equal volumes of the fluorochrome and algal sample are mixed and used under a fluorescent microscope. For picoplankton, a drop of the FDA is placed in a petridish and the processed nucleopore filter is transferred with the filtered surface up. The sample is allowed to stand for a few minutes and then processed further for picoplankton counts under a fluorescence microscope.

3.2. Computation of ecological indices from the plankton community

Whatever may be the sampling technique or the staining procedure, the main aim of any ecological study of plankton is to know their community structure. The community structure of plankton not only reflects their ambient water quality, but may also pinpoint the stress imparted by different environmental parameters (like salinity, sewage load, industrial discharge etc.) on them. A long term study may also lead to identification of opportunistic species or sensitive species in the planktonic community with reference to a particular variable or a group of variables.

The community structure of a planktonic community is determined through several ecological indices, but the first and foremost approach for such study is to know the population count of each species after proper taxonomic identification. This would lead to generate data on standing stock of planktonic community in a particular space at a particular time. However the stock assessment has to be done separately for phytoplankton and zooplankton to understand the grazing pressure or the production efficiency of each tier of the trophic level.

The various steps to evaluate the community structure of planktonic community are discussed here in brief:

Phase 1: Collection & concentration

In this step, collection and concentration of plankton is done to obtain a representative sample from a known volume of water. Conical nylon net bags (30 cm diameter) made of a 30 No. bolting silk cloth is used for filtering a known volume of water (may be 100 l or 200 l). The water samples collected within bucket of a known volume (say if it is a 5 litre bucket, it is filled 20 times to get 100 l of sample water) are filtered through the bolting silk cloth and the plankton is concentrated. Centrifugation is often done to concentrate the sample. The final volume of plankton concentrate is recorded to achieve the result of plankton density in terms of cells/ litre or cells/ m³. It is to be noted in this context that concentration by centrifugation often damages certain groups of zooplankton. This can be overcome by using a simple plankton concentrator that is quiet gentle in action. The concentrator consists of a stiff tube (12 cm diameter and 10 cm height) of perspex or PVC and a filter attached at the lower end. This filter may be a filter paper (Whatman No. 42) or a membrane filter supported by a monofilament nylon netting, which is attached with the lower end of the tube with ethylene dichloride (Fig. 3.2.1). The tube is dipped slowly into a 2 litre beaker containing the plankton sample. Water flows slowly upwards through the filter into the tube and is removed by a large pipette.

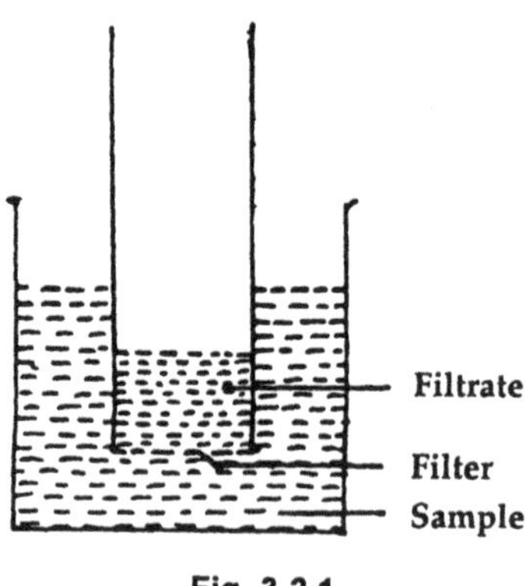

Fig. 3.2.1

With this type of concentrator, about 20 ml of zooplankton can be concentrated in about one hour.

Phase 2: Precounting processing

This step involves the counting of plankton through Sedgwick Rafter. One ml of plankton sample obtained from the stock through the pipette is transferred to the counting chamber – Sedgwick Rafter counting cell. The sample for

Fig. 3.2.2

counting in this chamber is spread evenly in the form of a thin layer and this is done by placing a cover slip diagonally across the counting cell and the sample is then introduced at one of its corner (Fig. 3.2.2). The cover slip moves into its proper position by capillary action. Sometimes the sample is too concentrated in terms of plankton density and under such circumstances, there is a probability of several planktonic organisms to be in a clustered form or to remain too close to one another. This makes the entire counting troublesome leading to inaccurate results. This is however avoided by diluting the stock plankton sample as per the requirement.

Phase 3: Counting

In this step, the process of counting is initiated by placing the Sedgwick Rafter counting cell (filled with plankton sample) under the microscope, which is provided with a mechanical stage. The plankton are counted individually from one corner of the counting cell. The Rafter is moved horizontally along the first row of the squares and the plankton in each square of the row are thus counted (sometimes species wise as per the requirement). After the completion of one row, the next adjacent row is adjusted using the mechanical device of the stage. In this way, all the plankton present in all the squares of the Sedgwick Rafter counting cell are recorded and counted. For every individual plankton species, a vertical stroke (I) is given. Thus for four separate individuals of the same species, four vertical strokes (IIII) are given in the record book. The fifth individual plankton, is however, represented by a diagonal stroke across the four vertical strokes (IIII). For statistical treatment or in order to increase the degree of accuracy, replicate counts of the one ml sample are done by carefully emptying, washing and successive filling of the counting chamber. The species documented through such process is finally recorded to enumerate the taxonomic diversity of plankton. At the end of this phase, a systematic chart of this format (Table 3.2.1) may be obtained from a given sample of water.

TABLE 3.2.1 : Species count of phytoplankton (A, B and C) and zooplankton (D, E and F) in 1 ml of sample water

Phytoplankton	
A	IIII
B	II
C	ℕℕ

Zooplankton	
D	ℕℕ, ℕℕ, IIII
E	IIII
F	II

The total number of plankton (standing crop) present in a litre of water sample can be calculated using the formula:

$$N = nv/V$$

Where,

N =   total number plankton cells per litre of water filtered.

n =   average number of plankton cells in 1 ml of plankton sample.

v =   volume of plankton concentrate (ml)

V =   volume of total water filtered (l)

The units of standing crop are N/l or $N \times 10^3/m^3$

Phase 4: Calculation of ecological indices

Ecological indices often reflect the degree of stress imposed on plankton community due to variation of environmental parameters. Several species of plankton are very sensitive to stress imparted by heavy metals, petroleum hydrocarbons or pesticides. This often creates a change in the community structure of the plankton, which is reflected through estimation of some common ecological indices as discussed here.

1. INDEX OF DOMINANCE

The index of dominance (Simpson, 1949) gives us an idea regarding the magnitude of stress in a particular community. It is calculated according to the following expression:

Index of dominance (c) $= \sum \left(\dfrac{n_i}{N}\right)^2$

Where, n_i = importance value for each species (number of individual, biomass, production etc.) and N= total of importance values.

The index of dominance is always lower where the dominancy is shared by a large number of species (Whittaker, 1965) or the total populations of the community is **uniformly** distributed among different species that is usually witnessed in clean, pollution free waters (Osborne *et al*, 1976) or in habitat having a unique physico-chemical variables. Infact, it is a general phenomenon that when the environmental stress in any particular habitat increases then the community in that habitat would be dominated by a fewer species (species that are more resistant to the existing set of physico-chemical and biological variables) which are often referred to as the **opportunistic species**. The magnitude of the index of dominance is usually high in stressful habitat. In the domain of phytoplankton community of Indian Sundarbans, the members of genus *Coscinodiscus* have been found to grab the major share through out the year.

2.   INDEX OF SIMILARITY

Two communities, geographically wide apart from each other may also show similarity in respect of their species composition. The similarity arises due to the closeness of the two habitats in respect of environmental characteristics.

The index of similarity between two sampling spots is calculated according to the following expression:

$$\textbf{Index of similarity (s)} = \frac{2C}{A+B}$$

Where, A = number of species present at sampling station "A", B = number of species present at sampling station "B" and C = number of species common to both the sampling stations.

3.   RICHNESS INDEX

The mathematical expressions of richness index are given below:

$$\textbf{Richness index } (d_i) = \frac{S-1}{\log_e N} \quad \text{(After Margalef, 1958)}$$

and

$$\textbf{Richness index (d}_2\textbf{)} = \frac{S}{\sqrt{N}} \quad \text{(After Menhinick, 1964)}$$

Where, S = number of species and N = total number of individuals of all the species.

The index reflects the suitability of a particular habitat for successful thriving and growth of different species. The value of the index usually decreases, when environment becomes unfavourable or stressful due to intrusion of some foreign matter e.g., pollutants or any other biological organisms that may be a parasite or predator to the existing group of species.

4. SHANNON WEINER SPECIES DIVERSITY INDEX

It is the most commonly used index in ecological studies owing to the fact that it is dimensionless, independent of sample size and expresses the worth of each species (Kochsiek $et\ al$, 1971). It is expressed as follows:

$$\textbf{Species diversity index } (\overline{\overline{H}}) = -\sum_{i=1}^{s} P_i \, log_e \, P_i$$

$$\text{or } (\overline{\overline{H}}) = -\sum_{i=1}^{s} \frac{n_i}{N} \, log_e \frac{n_i}{N}$$

where, P_i = importance probability for each species, n_i = importance value for each species and N = total of importance values.

The value of $\overline{H}$ has several ecological explanations. As Margalef (1968) stated that, " the ecologists find in any measure of diversity an expression of the possibilities of constructing feedback systems". Higher diversity, then, signifies longer food chains and more cases of symbiosis (mutualism, commensalism etc.) and greater probabilities for negative feedback control, which reduces the drastic oscillations and hence increases stability.

The value of $\overline{H}$ is also a unique indicator of environmental stress. As the sensitive species gradually shift or get eliminated from a habitat with the increase of the magnitude of environmental stress, therefore the species diversity index has been claimed as an effective statistics for predicting the change in environment (Wilhm and Dorris, 1968; Cairns and Dickson, 1971). Thus decrease in the value of $\overline{H}$ indicates an increase in the magnitude of environmental stress on the species, and the gradual restoration of the environmental quality is indicated by an increment in the value of $\bar{H}$ with the passage of time through the recruitment of new species.

5. EVENNESS INDEX

It is a measure of the uniformity of different species in a community and the value increases as the environment becomes favourable. The following expression is used to calculate the index:

$$\text{Evenness index (e)} = \frac{\bar{H}}{log_e\ S}$$

Where, $\bar{H}$ = Shannon Weiner species diversity index and S = number of species.

A very simplified programme has been developed here to compute these indices in PC (Programme I).

PROGRAM I

```
C   BIODIVERSITY INDEX
C
    DIMENSION A(1000)
    WRITE(*, 3)
3   FORMAT(8X, 'The season is : ')
    WRITE(*, *)
    WRITE(*, 4)
4   FORMAT(8X,'The Name of the Station : ')
    WRITE(*, *)
    WRITE(*, 8)
8   FORMAT(8X, 'Enter the No. of Species : ')
    WRITE(*, *)
    Read(*, 2)M
2   FORMAT(I5)
    WRITE(*, 9)
```

```
 9  FORMAT(8X, 'The No. of individuals are : ')
    WRITE(*, *)
    SUM=0.
    DO 30 I=1, M
    READ(*, *)A(I)
30  CONTINUE
    WRITE(*, *)
    WRITE(*, 40)M
40  FORMAT(8X, 'The No of Sp. = M= ', I5)
    WRITE(*, 60)
    DO 50 I=1, M
    WRITE(*, *)A(I)
50  CONTINUE
60  FORMAT(8X, 'The No. of individuals are : ')
    WRITE(*, *)
    DO 70  I=1, M
    SUM=SUM+A(I)
70  CONTINUE
    WRITE(*, 5)
 5  FORMAT(8X,  'Relative Abundance of the ith Sp. : ')
    DO 6 I=1, M
    R=A(I)*100/SUM
    WRITE(*, *)R
 6  CONTINUE
    WRITE(*, *)
    WRITE(*, 80)SUM
80  FORMAT(8X, 'Total No. of individuals of all species =N
    =', F10.4)
    BUM=0.
    TUM=0.
    DO 90 I=1, M
```

```
        X=A(I)/SUM
        D=X*X
        Y=ALOG(X)
        Z=X*Y
        BUM=BUM+Z
        TUM=TUM+D
90   CONTINUE
        H=-1* BUM
        WRITE(*, *)
        WRITE(*, 11)H
11   FORMAT(8X, 'Bio Diversity Index, H bar = ',F10.4)
        WRITE(*, *)
        E=1/H
        WRITE(*, 7)E
7    FORMAT (8X, 'The Equatibility Index, 1/H Bar =  ',F10.4)
        WRITE(*, *)
        P=H/ALOG(M)
        WRITE(*, 21)P
21   FORMAT(8X, 'Evenness index = H bar / Ln (M) = ',F10.4)
        B=SQRT(SUM)
        C=M/B
        WRITE(*, *)
        WRITE(*, 31)C
31   FORMAT(8X, 'Richness index = S/ SQRT (N) = ',F10.4)
        WRITE(*, *)
        WRITE(*, 41)TUM
41   FORMAT(8X, 'Index of Dominance = ',F10.4)
        WRITE(*, *)
        WRITE(*, *)
        STOP
        END
```

Suggested references

1. Bentley – Mowat, J. (1982). Application of fluorescence microscopy to pollution studies on marine phytoplankton. Bot. Mar. 25: 203-204.

2. De Noyelles, F. (Jr) (1968). A stained – organism filter technique for concentrating phytoplankton. Limnol. Oceanogr. 13: 562-565.

3. Reynolds, A. E., Mackiernan, G. B. and Valkenburg, S. D. V. (1978). Vital and mortal staining of algae in the presence of chlorine-produced oxidants. Estuaries 1: 192-196.

4. Simpson, E.H. (1949). Measurement of diversity. Nature, 163, 688.

5. Osborne, A., Wanielista, P. and Yousef, A. (1976). Benthic fauna species diversity in six Central Florida lakes in summer. Hydrobiologia, 48, 125-129.

6. Kochsiek, K. A., Wilhm, J. L. & Morison, R. (1971). Species diversity of net zooplankton and physico-chemical conditions in keystone reservoir, Oklahoma. Ecology, 52, 1119-1125.

7. Margalef, R. (1968). Perspectives in Ecological Theory. University of Chicago Press, Chicago.

8. Cairns, J. Jr and Dickson, K.L. (1971). A simple method for the biological assessment of the effects of waste discharges on aquatic bottom-dwelling organisms. Journal of Water Pollution Control Federation. 43, 755-772.

9. Shannon, C.E. and Weiner, V. (1949). The Mathematical theory of Communications, University of Illionis Press, Urbana. 117 p.

Internet references

- *http://globec.whoi.edu/globec_dir/ICES_CD/phyto/methods.htm*
- *http://www.cals.ncsu.edu/course/zo419/phytotan.htm*

Chapter 4
PLANKTON SPECTRUM OF NORTHWESTERN BAY OF BENGAL

4.1. Common phytoplankton of northwestern Bay of Bengal

Phytoplankton are the primary producers of organic food supply in the marine food chain and therefore serve as important basic link in the production and sustenance of life at sea, including fish. The northwestern Bay of Bengal has a nutrient rich coastal waters owing to the presence of Sundarbans mangrove ecosystem at the apex. The detritus supplied by these halophytic vegetations saturate the adjacent waterbodies with nutrients, due to which a rich diversity in the phytoplankton community has been observed (Mitra and Pal, 2002). Studies on species composition, abundance and distribution of phytoplankton identified 35 species of phytoplankton (Chaudhuri and Choudhury, 1994). Later on some more species were recorded and the number increased to 48 (Mitra and Choudhury, 2000). The present authors documented 102 phytoplankton species from 20 different stations of coastal West Bengal (Fig. 1-102). The *Coscinodiscus* spp. dominated the composition. The species spectrum of phytoplankton in and around north western Bay of Bengal, as documented by the present authors is highlighted in Table 4.1.1.

TABLE 4.1.1 : Phytoplankton spectrum around Northwestern Bay of Bengal coast

S.No.	Name	Systematic position	Salient features
1.	*Coscinodiscus eccentricus*	Division : Thallophyta Class:Bacillariophyceae Order : Centrales Sub-order: Coscinodiscineae Family : Coscinodisceae Genus : *Coscinodiscus* Species : *eccentricus*	• The cells are disc shaped with two valves - epitheca and hypotheca. • Cells are double walled with hexagonal markings and areolae of same size arranged in tangential series. • Diameter of the cell is 34 to 104 µm.
2.	*Coscinodiscus jonesianus*	Division: Thallophyta Class: Bacillariophyceae Order : Centrales Suborder: Coscinodiscineae Family : Coscinodisceae Genus : *Coscinodiscus* Species : *jonesianus*	• The cells are normally large with 4 central rosette of areolae. • Presence of radial and spiral rows of areolae from margin towards centre. • Diameter of the cell is 140 to 215 µm.
3.	*Coscinodiscus lineatus*	Division: Thallophyta Class: Bacillariophyceae Order : Centrales Suborder: Coscinodiscineae Family : Coscinodisceae Genus : *Coscinodiscus* Species : *lineatus*	• Cells are normally disc shaped with linearly arranged cells. • Cells have narrow differentiated margins. • Diameter of the cell is 30 to 100 µm.
4.	*Coscinodiscus radiatus*	Division:Thallophyta Class: Bacillariophyceae Order : Centrales Suborder: Coscinodiscineae Family : Coscinodisceae Genus : Coscinodiscus Species : radiatus	• Cells are disc shaped and arranged radially from centre to periphery. • This species is comparatively larger in size than *C. eccentricus* & *C. lineatus*. • Diameter of the cell is 460 to 530 µm.

S.No.	Name	Systematic position	Salient features
5.	*Coscinodiscus gigas* 	Division : Thallophyta Class: Bacillariophyceae Order : Centrales Suborder: Coscinodiscineae Family : Coscinodisceae Genus : *Coscinodiscus* Species : *gigas*	• The cells are large and flat very similar to a coin. • Small areolae are found at the margins with larger areolae at the centre and a central large area. • Diameter of the cell is 472 to 536 µm.
6.	*Coscinodiscus oculus-iridis* 	Division : Thallophyta Class: Bacillariophyceae Order : Centrales Suborder: Coscinodiscineae Family : Coscinodisceae Genus : *Coscinodiscus* Species : *oculus-iridis*	• The cells are large and disc shaped. • Areolae increase slightly towards margin with the margin small and radially striated. • Diameter of the cell is 155 to 172 µm.
7.	*Coscinodiscus concinnus* 	Division : Thallophyta Class: Bacillariophyceae Order : Centrales Suborder : Coscinodiscineae Family : Coscinodisceae Genus : *Coscinodiscus* Species : *concinnus*	• The cells are large with areolae in the centre becoming smaller and slender. • Presence of about 8-10 areolae in 10 µm at the centre and 11-12 in 10 µm near the margin. • Diameter of the valve is 250 to 300 µm.
8.	*Coscinodiscus perforatus* 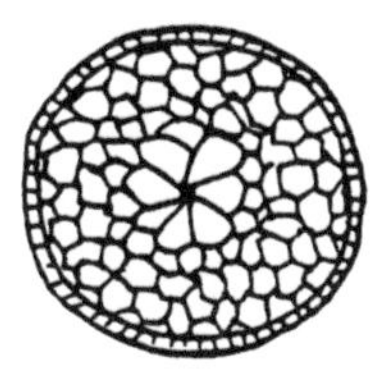	Division : Thallophyta Class: Bacillariophyceae Order : Centrales Suborder: Coscinodiscineae Family : Coscinodisceae Genus : *Coscinodiscus* Species : *perforatus*	• The cells are disc shaped with areolae arranged in a central rosette in radial rows. • Areolae 3-5 in 10 µm around the rosette and 3-4 in 10 µm near the margin. • Valves 145 to 165 µm diameter and margin is striated.

S.No.	Name	Systematic position	Salient features
9.	*Coscinodiscus asteromphalus* 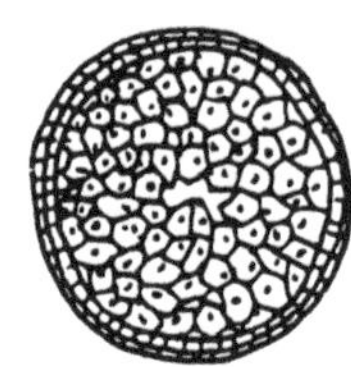	Division : Thallophyta Class: Bacillariophyceae Order : Centrales Suborder: Coscinodiscineae Family : Coscinodisceae Genus : *Coscinodiscus* Species: *asteromphalus*	• Valves depressed at the centre. • Areolae 3-4 in 10 µm near centres and 4-5 in 10 µm near the margin. • Valves are 200 to 215 µm in diameter.
10.	*Coscinodiscus thorii* 	Division : Thallophyta Class: Bacillariophyceae Order : Centrales Suborder: Coscinodiscineae Family : Coscinodisceae Genus : *Coscinodiscus* Species : *thorii*	• Cells are disc shaped, areolae are partially seen, smaller in size and arranged in tangential series. • Presence of chromatophores is the characteristic of the cell. • Valve diameter is 160 µm.
11.	*Cyclotella* sp. 	Division : Thallophyta Class: Bacillariophyceae Order : Centrales Family : Coscinodisceae Genus : *Cyclotella* sp.	• Cells are disc shaped. • Regular striations extend from centre towards periphery. • Diameter of the cell is 35 to 50 µm.
12.	*Cyclotella striata* 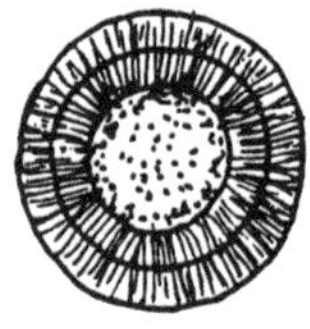	Division : Thallophyta Class: Bacillariophyceae Order : Centrales Family : Coscinodisceae Genus : *Cyclotella* Species : *striata*	• Cells are disc shaped with central area coarsely punctae. • There are numerous regular striations with evenly striated border. • Diameter of the cell is 15 to 30 µm.

S.No.	Name	Systematic position	Salient features
13.	*Thalassiosira subtilis* 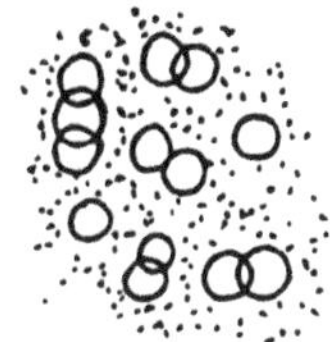	Division : Thallophyta Class: Bacillariophyceae Order : Centrales Family : Coscinodisceae Genus : *Thalassiosira* Species : *subtilis*	• The cells are mostly small and disc shaped. • These are normally found in colonies embedded in mucilage. • Diameter of the cell is 36 to 41 µm.
14.	*Thalassiosira* sp. 	Division : Thallophyta Class: Bacillariophyceae Order : Centrales Family : Coscinodisceae Genus: *Thalassiosira* sp.	• The cells are normally disc shaped and are either free living or colonial. • The chromatophores are scattered and numerous in the cell. • Diameter of the cell is 40 to 50 µm.
15.	*Skeletonema costatum* 	Division : Thallophyta Class: Bacillariophyceae Order : Centrales Family : Coscinodisceae Genus : *Skeletonema* Species : *costatum*	• The cells form long and slender chains with the help of marginal spines. • The valves are small, lens shaped and space between the cells is larger than the cells. • Diameter of the cell is 12 to 15 µm.
16.	*Paralia sulcata* 	Division : Thallophyta Class: Bacillariophyceae Order : Centrales Family : Coscinodisceae Genus : *Paralia* Species : *sulcata*	• The cells are disc shaped with 3 outer layers. • Presence of teeth like intrusions in the innermost layers. • Central core area is coarse with the cell diameter 445-520 µm.

S.No.	Name	Systematic position	Salient features
17.	*Planktoniella sol* 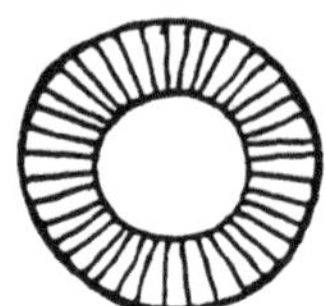	Division : Thallophyta Class: Bacillariophyceae Order : Centrales Family : Coscinodisceae Genus : *Planktoniella* Species : *sol*	• The cells are disc shaped, small with flat valves. • Presence of a characteristic wing like expansion with weakly silicified and radial rays hanging from the epitheca. • Diameter of valve is 65-73 µm.
18.	*Rhizosolenia crassispina* 	Division : Thallophyta Class :Bacillariophyceae Order : Centrales Family : Soleniae Genus : *Rhizosolenia* Species : *crassispina*	• Cells are cylindrical with truncated ends. • Apical processes with hair-like aftergrowths. • Diameter and length of the cells are 41-54 µm and 145-278 µm respectively.
19.	*Rhizosolenia setigera* 	Division : Thallophyta Class :Bacillariophyceae Order : Centrales Family : Soleniae Genus : *Rhizosolenia* Species : *setigera*	• Cells are rod shaped and cylindrical. • Apical processes long and end in a spine. • Diameter of the cell is 7-8 µm and length is 512-528 µm.

S.No.	Name	Systematic position	Salient features
20.	*Rhizosolenia alata* 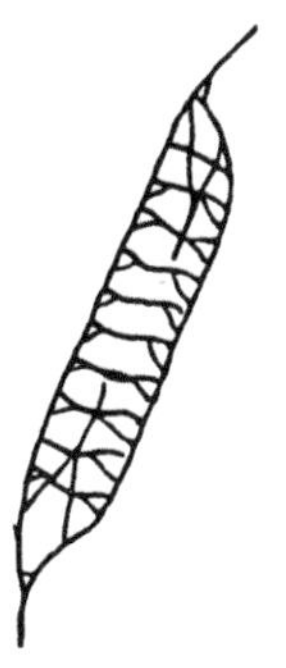	Division : Thallophyta Class :Bacillariophyceae Order : Centrales Family : Soleniae Genus : *Rhizosolenia* Species : *alata*	• The cells are rod shaped, cylindrical with curved tube like processes. • The depression of each process fits the adjoining cell. • Diameter is 6-31 µm and length is 412-668 µm.
21.	*Rhizosolenia hebetata*	Division : Thallophyta Class :Bacillariophyceae Order : Centrales Family : Soleniae Genus : *Rhizosolenia* Species : *hebetata*	• Cells are similar to *R. setigera*. • Apical processes is long, hollow and ends in a long spine. • Diameter is 6-15 µm and length is 218-588 µm.
22.	*Rhizosolenia styliformis*	Division : Thallophyta Class :Bacillariophyceae Order : Centrales Family : Soleniae Genus : *Rhizosolenia* Species : *styliformis*	• Cells are cylindrical with spines long and hollow. • *R. intracellularis* found inside the cell. • Diameter is 21-101 µm and length is 210-385 µm.

S.No.	Name	Systematic position	Salient features
23.	*Rhizosolenia robusta*	Division : Thallophyta Class :Bacillariophyceae Order : Centrales Family : Soleniae Genus : *Rhizosolenia* Species : *robusta*	• Cells are crescent shaped and somewhat cylindrical in the middle with diameter of 51-274 µm. • Valve ends curved, tapering with a small spine set in hollow apical process of valve. • Presence of numerous intercalary bands.
24.	*Rhizosolenia stolterfothii*	Division : Thallophyta Class :Bacillariophyceae Order : Centrales Family : Soleniae Genus : *Rhizosolenia* Species : *stolterfothii*	• Cells are cylindrical and form spirally coiled chains. • Each valve possess spine and numerous chromatophores. • Diameter is 18-36 µm and length is 120-165 µm.
25.	*Lauderia annulata*	Division : Thallophyta Class :Bacillariophyceae Order : Centrales Family : Soleniae Genus : *Lauderia* Species : *annulata*	• Cells are cylindrical and measure 32-50 µm long and 55-70 µm diameter. • Valves convex with a depression in the middle, forming a straight chain and raised at margin touching the adjacent cell. • Presence of numerous short spines of varying length in the valves.

S.No.	Name	Systematic position	Salient features
26.	*Bacteriastrum delicatulum* 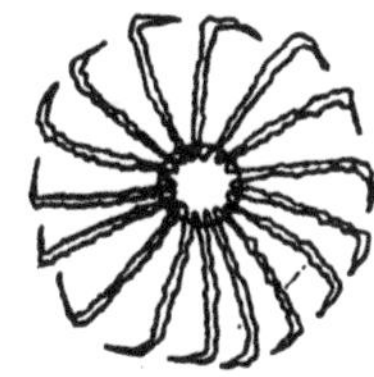	Division : Thallophyta Class: Bacillariophyceae Order : Centrales Family : Chaetocereae Genus : *Bacteriastrum* Species : *delicatulum*	• Cells are longer than broad with setae perpendicular to chain axis and are 6 to 12 in number. • Setae of the two adjoining cells fuse at the base and go apart as branches. • Diameter of the cell varies from 6 to 12 µm.
27.	*Bacteriastrum varians* 	Division : Thallophyta Class :Bacillariophyceae Order : Centrales Family : Chaetocereae Genus : *Bacteriastrum* Species : *varians*	• Cells form chains with 8 to 19 setae arranged at right angles to chain axis. • Setae extend from chain axis with fine spines arranged in spiral rows. • Diameter of the cell is 13 to 38 µm.
28.	*Bacteriastrum comosum* 	Division : Thallophyta Class: Bacillariophyceae Order : Centrales Family : Chaetocereae Genus : *Bacteriastrum* Species : *comosum*	• Presence of semicircular knob like structure in between each valve. • Setae extend from the knob and bend in a clockwise manner, hanging downward. • Setae are 6-11 which fuse with the adjoining cells.
29.	*Bacteriastrum hyalinum*	Division : Thallophyta Class: Bacillariophyceae Order : Centrales Family : Chaetocereae Genus : *Bacteriastrum* Species : *hyalinum*	• Cells form chains which are broader than long. • Setae many (12-25) and perpendicular to chain axis which fuse at the base and go apart as branches. • Diameter 32-38 µm.

S.No.	Name	Systematic position	Salient features
30.	*Chaetoceros dydymus* 	Division : Thallophyta Class: Bacillariophyceae Order : Centrales Family : Chaetocereae Genus : *Chaetoceros* Species : *dydymus*	• Cells form straight chains with semicircular knob in between each valve. • Distinct interlocking of setae are seen. • Length of the cell is 22-40 µm.
31.	*Chaetoceros curvisetus*	Division : Thallophyta Class: Bacillariophyceae Order : Centrales Family : Chaetocereae Genus : *Chaetoceros* Species : *curvisetus*	• Cells form curved long chains. • All the setae are directed towards one side of the chain. • Length of the cell is 8-20 µm.
32.	*Chaetoceros diversus*	Division : Thallophyta Class: Bacillariophyceae Order : Centrales Family : Chaetocereae Genus : *Chaetoceros* Species : *diversus*	• The cells are compact with small apertures. • Setae of cells are straight, thicker, tubular and spinous. • Length of the cell is 5-9 µm.
33.	*Chaetoceros messanensis*	Division : Thallophyta Class :Bacillariophyceae Order : Centrales Family : Chaetocereae Genus : *Chaetoceros* Species : *messanensis*	• Cells form long and straight chains with apertures round or ovate. • Some of the inner setae are closely placed and bifurcating farther. • Length of the cell is 12-41 µm.

S.No.	Name	Systematic position	Salient features
34.	*Chaetoceros peruvianus*	Division : Thallophyta Class :Bacillariophyceae Order : Centrales Family : Chaetocereae Genus : *Chaetoceros* Species : *peruvianus*	• Cells are found solitary. • Setae of upper valve starts from centre of the valve with both anterior and posterior setae long and parallel to perivalvar axis. • Length of the cell is 8-32 µm.
35.	*Chaetoceros eibenii*	Division : Thallophyta Class :Bacillariophyceae Order : Centrales Family : Chaetocereae Genus : *Chaetoceros* Species : *eibenii*	• Cells cylindrical forming chains with apertures of adjoining cells elliptical. • Setae with bulbous base and armed with spines distally. • Frustules 23-28 µm long and 30-325 µm diameter.
36.	*Chaetoceros lorenzianus*	Division : Thallophyta Class :Bacillariophyceae Order : Centrales Family : Chaetocereae Genus : *Chaetoceros* Species : *lorenzianus*	• Valve surface slightly elevated at the centre. • Setae four sided with terminal setae thicker and inner setae interlocking. • Frustules 25-40 µm long and 45-75 µm diameter.
37.	*Chaetoceros indicus* 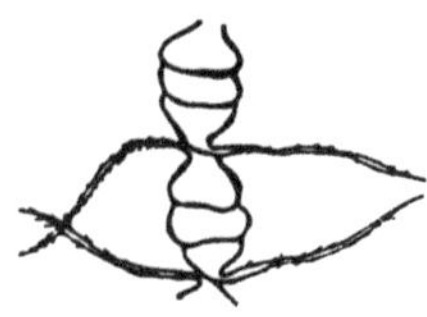	Division : Thallophyta Class :Bacillariophyceae Order : Centrales Family : Chaetocereae Genus : *Chaetoceros* Species : *indicus*	• Cells form straight chains and appear sphere shaped in girdle view. • Setae start from corners armoured with short spines. • Frustules 22-28 µm long 15-17 µm diameter.

S.No.	Name	Systematic position	Salient features
46.	*Biddulphia sinensis* 	Division : Thallophyta Class: Bacillariophyceae Order : Centrales Family : Biddulphieae Genus : *Biddulphia* Species : *sinensis*	• Cells form short chains with cylindrical to square in girdle view. • Presence of two thin blunt horns with two long and thin spines. • Length of the cell is 82-215 µm.
47.	*Biddulphia mobiliensis* 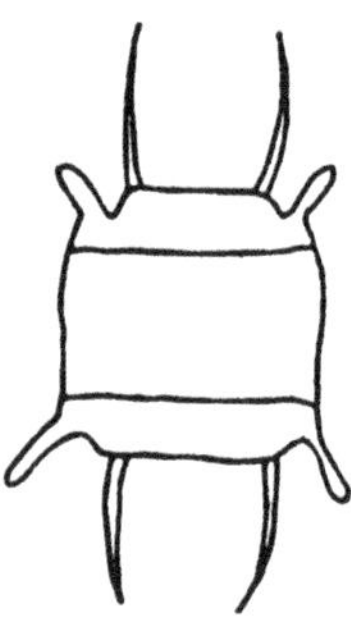	Division : Thallophyta Class: Bacillariophyceae Order : Centrales Family : Biddulphieae Genus : *Biddulphia* Species : *mobiliensis*	• It is almost similar in shape as *B. sinensis* but a little more flattened. • Presence of four thin blunt horns at the corners of the valve with four thin long spines. • Length of the cell is 24-81 µm.
48.	*Eucampia zoodiacus* 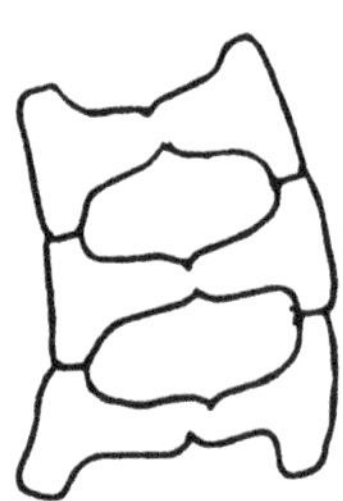	Division : Thallophyta Class: Bacillariophyceae Order : Centrales Family : Biddulphieae Genus : *Eucampia* Species : *zoodiacus*	• Cells flat, united, form spirally twisted chains with blunt processes. • Valves concave in middle with wide aperature between two cells. • Length of the cell is 43-61 µm.

S.No.	Name	Systematic position	Salient features
49.	*Hemidiscus hardmannianus*	Division : Thallophyta Class: Bacillariophyceae Order : Centrales Family : Euodieae Genus : *Hemidiscus* Species: *hardmannianus*	• Valves are semicircular with central area large and hyaline. • Ventral margins are more or less straight. • Fine areolation are seen radiating from the centre.
50.	*Hemidiscus cuneiformis*	Division : Thallophyta Class: Bacillariophyceae Order : Centrales Family : Euodieae Genus : *Hemidiscus* Species : *cuneiformis*	• Valves are semicircular. • Ventral margins straight and ends obtuse. • Frustules 300-400 µm diameter and valves 140-200 µm broad.
51.	*Climacosphenia elongata*	Division : Thallophyta Class: Bacillariophyceae Order : Pennales Family : Fragilarioideae Genus : *Climacosphenia* Species : *elongata*	• Cells are cuneate and rounded at angles with truncate base. • Cells are elongated and become linear. • Fine striae 21-25 in 10 µm; length 710-790 µm; breadth 26-28 µm and 9 µm at base.
52.	*Fragilaria oceanica*	Division : Thallophyta Class: Bacillariophyceae Order : Pennales Family : Fragilarioideae Genus : *Fragilaria* Species : *oceanica*	• Cells are rectangular in girdle view with valves lanceolate & pseudoraphe narrow and linear. • Transapical striae 14 in 10 µm. • Length of cell is 11-32 µm and breadth being 6 µm.

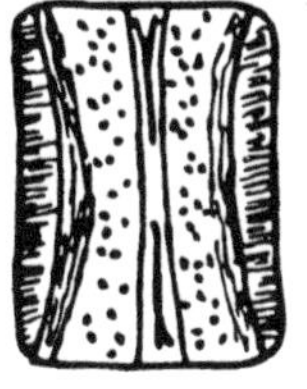

S.No.	Name	Systematic position	Salient features
53.	*Rhaphoneis amphiceros* 	Division : Thallophyta Class: Bacillariophyceae Order : Pennales Family : Fragilarioideae Genus : *Rhaphoneis* Species : *amphiceros*	• Cells are spindle shaped with both ends tapering. • Usually solitary cells, pseudoraphe narrow with transapical striae 9 in 10 µm. • Length is 25-41 µm and breadth is 16-24 µm.
54.	*Thalassionema nitzschioides* 	Division : Thallophyta Class: Bacillariophyceae Order : Pennales Family : Fragilarioideae Genus : *Thalassionema* Species : *nitzschioides*	• Cells form zig-zag chains and are usually rectangular in shape. • Cells rest at protoplasmic cushions found at junctions. • Marginal striae 12 in 10 µm; with the length of cell being 20-66 µm and breadth is about 3 µm.
55.	*Thalassiothrix longissima* 	Division : Thallophyta Class: Bacillariophyceae Order : Pennales Family : Fragilarioideae Genus : *Thalassiothrix* Species : *longissima*	• These are solitary cells and thread-like in appearance. • Valves are linear with rounded ends. • Length of the cell is 495-1750 µm and breadth is 2.5 µm.
56.	*Thalassiothrix fraunfeldii* 	Division : Thallophyta Class: Bacillariophyceae Order : Pennales Family : Fragilarioideae Genus : *Thalassiothrix* Species : *fraunfeldii*	• Cells form zig-zag or star like chains. • The base of the cells rest at protoplasmic cushions with their poles separate. • Marginal striae 12 in 10 µm with a cell length of 96-235 µm.

S.No.	Name	Systematic position	Salient features
57.	*Thalassiothrix nitzschioides* 	Division : Thallophyta Class: Bacillariophyceae Order : Pennales Family : Fragilarioideae Genus : *Thalassiothrix* Species : *nitzschioides*	• Cells form zig-zag chains. • Usually smaller in size with poles distinct and dissimilar. • Length of the cell is 25-70 µm.
58.	*Asterionella japonica* 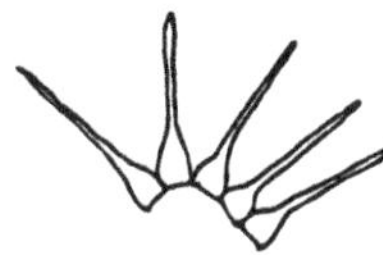	Division : Thallophyta Class: Bacillariophyceae Order : Pennales Family : Fragilarioideae Genus : *Asterionella* Species : *japonica*	• Cells form spiral colonies. • Cells are knob like at base with slender long apex. • Length of the cell is 43-106 µm and breadth is 7-11 µm.
59.	*Diatoma vulgare* 	Division : Thallophyta Class: Bacillariophyceae Order : Pennales Family : Fragilarioideae Genus : *Diatoma* Species : *vulgare*	• Valves lanceolate. • Transverse coastae present across the valve. • Frustules 28-33 µm long and 5.5-6.3 µm diameter.
60.	*Cocconeis* sp. 	Division : Thallophyta Class: Bacillariophyceae Order : Pennales Family: Achnanthoideae Genus : *Cocconeis* sp.	• Cells are ovate with three hyaline areas. • Striae unequal in length varying about 20 in 10 µm with sigmoid raphe. • Length of the cell is 43-106 µm and breadth 7-11 µm.

S.No.	Name	Systematic position	Salient features
61.	*Acnanthus stromii* 	Division : Thallophyta Class: Bacillariophyceae Order : Pennales Family: Achnanthoideae Genus : *Acnanthus* Species: *stromii*	• Valves lanceolate with pseudoraphe, both transapical and longitudinal ribs. • Valves with narrow axial areas and widened in the middle. • Length of the cell is 34-42 µm and breadth 12-16 µm.
62.	*Gyrosigma balticum* 	Division : Thallophyta Class: Bacillariophyceae Order : Pennales Family : Naviculoideae Genus : *Gyrosigma* Species : *balticum*	• Valves are linear with truncated ends. • Raphe excentric with central small area. • Length and breadth of the cell are 290-338 µm and 28-30 µm respectively.
63.	*Pleurosigma normanii* 	Division : Thallophyta Class: Bacillariophyceae Order : Pennales Family : Naviculoideae Genus : *Pleurosigma* Species : *normanii*	• Valves are broadly lanceolate. • Ends are blunt with sigmoid raphe. • Length and breadth of the cell vary between 195-328 µm and 40-62 µm respectively.
64.	*Pleurosigma elongatum* 	Division : Thallophyta Class: Bacillariophyceae Order : Pennales Family : Naviculoideae Genus : *Pleurosigma* Species : *elongatum*	• Valves are elongated and sigmoid. • Ends are acute with sigmoid raphe, central in position. • Length is 205-398 µm and breadth is 32-40 µm.

S.No.	Name	Systematic position	Salient features
65.	*Diploneis smithii* 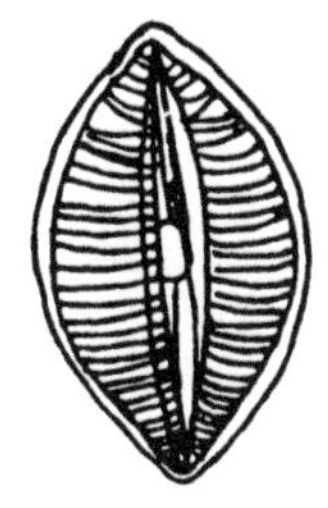	Division : Thallophyta Class: Bacillariophyceae Order : Pennales Family : Naviculoideae Genus : *Diploneis* Species : *smithii*	• The valves are ovate with rounded poles and central small quadrate nodule. • Raphe lanceolate with transapical costae 9 in 10 µm and radial with alternating rows of alveoli in two oblique rows. • Length is 55-60 µm and breadth is 35-38 µm.
66.	*Diploneis robustus* 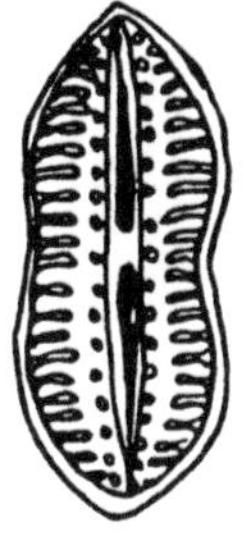	Division : Thallophyta Class: Bacillariophyceae Order : Pennales Family : Naviculoideae Genus : *Diploneis* Species : *robustus*	• Valves are fairly ovate with a middle constriction and rounded poles. • Transapical costae are well developed, swollen at tips with 6 in 10 µm. • Two rows of alveoli on either side of raphe but interrupted in middle with a cell length of 58-74 µm and breadth of 21-30 µm.
67.	*Navicula longa* 	Division : Thallophyta Class: Bacillariophyceae Order : Pennales Family : Naviculoideae Genus : *Navicula* Species : *longa*	• Valves are rhombic with pointed ends. • Central area small and axial area narrow. • Striae 9-11 in 10 µm with length and breadth being 52-56 µm and 10 µm respectively.

S.No.	Name	Systematic position	Salient features
68.	*Cymbella marina*	Division : Thallophyta Class: Bacillariophyceae Order : Pennales Family : Naviculoideae Genus : *Cymbella* Species : *marina*	• Cells are spindle shaped with raphe slightly broad. • Axial area narrow, but central area rather large with striae radial and 15 in 10 µm. • Length and breadth of the cell are 45-85 µm and 14-16 µm respectively.
69.	*Nitzschia sigma*	Division : Thallophyta Class: Bacillariophyceae Order : Pennales Family : Naviculoideae Genus : *Nitzschia* Species : *sigma*	• Valves are linear, somewhat sigmoid with a central bulge. • The two poles are very much pointed with punctae 5-6 in 10 µm. • Length of the cell is 280-310 µm.
70.	*Nitzschia longissima*	Division : Thallophyta Class: Bacillariophyceae Order : Pennales Family : Naviculoideae Genus : *Nitzschia* Species : *longissima*	• Valves are straight, hair like with middle portion lanceolate. • Keel punctae 12 in 10 µm but the striae are not visible. • Length of the cell is 72-460 µm and the breadth being 3-5 µm.

S.No.	Name	Systematic position	Salient features
71.	*Nitzschia closterium*	Division : Thallophyta Class: Bacillariophyceae Order : Pennales Family : Naviculoideae Genus : *Nitzschia* Species : *closterium*	• Valves are spindle shaped in middle with ends beak like. • Striae are not visible. • Length of the cell is 140-160 µm and breadth is 3-7 µm respectively.
72.	*Nitzschia striata*	Division : Thallophyta Class: Bacillariophyceae Order : Pennales Family : Naviculoideae Genus : *Nitzschia* Species : *striata*	• Cells are spindle shaped with pointed ends and are usually found in chains. • Ends of the cells are pressed against each other for a short distance. • Length is 45-128 µm and breadth is 3-5 µm respectively.
73.	*Pinnularia alpina*	Division : Thallophyta Class: Bacillariophyceae Order : Pennales Family : Naviculoideae Genus : *Pinnularia* Species : *alpina*	• Valves lanceolate and elliptic with fairly rounded ends. • Axial area large and striae broad and radiate, 4 in 10 µm. • Length is 80-84 µm and breadth is 27-28 µm.
74.	*Bacillaria paradoxa*	Division : Thallophyta Class: Bacillariophyceae Order : Pennales Family : Naviculoideae Genus : *Bacillaria* Species : *paradoxa*	• Valves linear, rectangular in shape to form mat like chain. • Transapical striae 20-22 in 10 µm. • Length is 108-202 µm and breadth is 6-9 µm.

S.No.	Name	Systematic position	Salient features
75.	*Prorocentrum gracile* 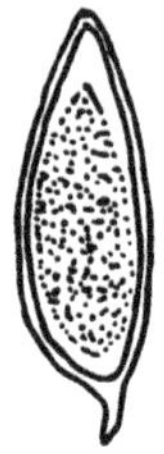	Division : Thallophyta Class : Pyrrophyceae Order : Prorocentrales Family: Prorocentraceae Genus : *Prorocentrum* Species : *gracile*	• Cells are elongated, cylindrical, rounded anteriorly and tapering posteriorly. • Presence of a slight sigmoid anterior process. • Length of the cell is 65 µm and breadth is 16 µm respectively.
76.	*Prorocentrum* sp. 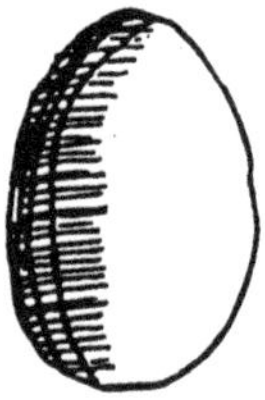	Division : Thallophyta Class : Pyrrophyceae Order : Prorocentrales Family: Prorocentraceae Genus:*Prorocentrum* sp.	• Cells are ovate, composed of two valves. • Longitudinal valves connected by suture and intercalary bands. • Chromatophores small and yellowish brown.
77.	*Ceratium furca* 	Division : Thallophyta Class : Pyrrophyceae Order : Peridiniales Family: Ceratiaceae Genus : *Ceratium* Species : *furca*	• Epitheca gradually narrowing to a short, long apical horn. • Left antapical longer and straight than right and pointed. • Dimension is 150-155 µm by 45-55 µm.
78.	*Ceratium teres* 	Division : Thallophyta Class : Pyrrophyceae Order : Peridiniales Family: Ceratiaceae Genus : *Ceratium* Species : *teres*	• Body is small with epitheca three-cornered and long, straight apical horn. • Hypotheca narrow with antapicals diverging, left slightly longer than right. • Dimension of the cell is 145-155 µm by 28-32 µm.

S.No.	Name	Systematic position	Salient features
79.	*Ceratium minutum* 	Division : Thallophyta Class : Pyrrophyceae Order : Peridiniales Family: Ceratiaceae Genus : *Ceratium* Species : *minutum*	• Cells are small with rounded body. • Anterior antapical horns are short. • Dimension of the cell is 60-70 µm by 15-20 µm respectively.
80.	*Ceratium extensum* 	Division : Thallophyta Class : Pyrrophyceae Order : Peridiniales Family: Ceratiaceae Genus : *Ceratium* Species : *extensum*	• Epitheca long, narrow, straight with narrow apical horn. • Left antapical straight, long, bent slightly and right antapical absent. • Very long body measures about 155-180 µm by 10-12 µm.
81.	*Ceratium tripos* 	Division : Thallophyta Class : Pyrrophyceae Order : Peridiniales Family: Ceratiaceae Genus : *Ceratium* Species : *tripos*	• Cell large, as long as broad with epitheca twice as broad as long. • Right horn weakly developed than left horn, but both horns are robust and equal. • Cell measures about 80-85 µm by 75-82 µm.
82.	*Ceratium trichoceros* 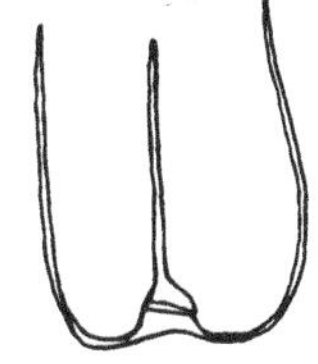	Division : Thallophyta Class : Pyrrophyceae Order : Peridiniales Family: Ceratiaceae Genus : *Ceratium* Species : *trichoceros*	• Very small body with long and thin horns. • Ends of antapicals parallel to apical horn. • Cell measures about 120-155 µm by 15-30 µm.

S.No.	Name	Systematic position	Salient features
83.	*Ceratium horridum*	Division : Thallophyta Class : Pyrrophyceae Order : Peridiniales Family: Ceratiaceae Genus : *Ceratium* Species : *horridum*	• Epitheca triangular in outline and shorter than hypotheca. • Left posterior horn converging and right posterior horn diverging from apical horn. • Cell proper measures about 39-49 µm by 38-42 µm.
84.	*Ceratium kofoidii*	Division : Thallophyta Class : Pyrrophyceae Order : Peridiniales Family: Ceratiaceae Genus : *Ceratium* Species : *kofoidii*	• Smaller species, epitheca triangular with thin apical horn. • Antapicals unequal, very thin and each ending in a long fine point, left antapical longer than right. • Cell measures 115-125 µm by 18-20 µm.
85.	*Ceratium falcatum*	Division : Thallophyta Class : Pyrrophyceae Order : Peridiniales Family: Ceratiaceae Genus : *Ceratium* Species : *falcatum*	• Apical horn straight and longer than antapical. • Antapical deflected towards left. • Cell measures 130-160 µm by 15-20 µm.
86.	*Ceratium inflatum*	Division : Thallophyta Class : Pyrrophyceae Order : Peridiniales Family: Ceratiaceae Genus : *Ceratium* Species : *inflatum*	• Epitheca slightly broad ending with apical horn sharp, tapering and bent. • Left antapical very long and sharp while right antapical very small. • Cell measures 145-160 µm by 15 µm.

S.No.	Name	Systematic position	Salient features
95.	*Oscillatoria limosa*	Division : Thallophyta Class : Cyanophyceae Genus : *Oscillatoria* Species: *limosa*	• Trichomes more or less straight, not constricted at cross walls. • End-cells flatly rounded with slightly thickened membrane. • Cell measures 11-14 µm broad and 2-4 µm long.
96.	*Dicrateria gilva*	Division : Thallophyta Class : Haptophyceae Genus : *Dicrateria* Species : *gilva*	• Cells are rounded and biflagellate. • These are unicellular and motile. • Number of chloroplasts varies from 1 to 4.
97.	*Phaeocystis* sp.	Division : Thallophyta Class : Haptophyceae Genus : *Phaeocystis* sp.	• Cells are four lobed. • Chloroplasts are scattered within the cell. • These are golden-brown in colour.
98.	*Dictyocha* sp.	Division : Thallophyta Class : Chrysophyceae Genus : *Dictyocha* sp.	• The cell is quadriflagellate, with three pseudoflagella and one flagella. • Presence of four chromatophores. • These are motile in nature, and possess internal skeleton of conspicous siliceous spicules.

S.No.	Name	Systematic position	Salient features
99.	*Tetraselmis gracilis* 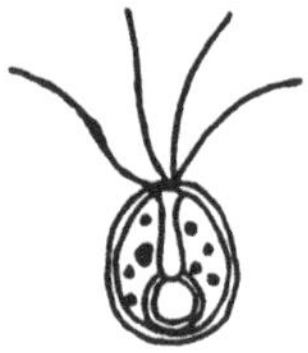	Division : Thallophyta Class : Prasinophyceae Genus : *Tetraselmis* Species : *gracilis*	• These are motile with four equal flagella. • Size of this organism is around 2 μm. • Cell consists of a single chloroplasts, pyrenoid, and protein coat.
100.	*Tetraselmis* sp. 	Division : Thallophyta Class : Prasinophyceae Genus : *Tetraselmis* sp.	• Cell is oval in appearance and is motile. • Presence of two equal size flagella anteriorly and two long flagella posteriorly emerging from a common point. • Cell structure not distinct with one chloroplast.
101.	*Chlorella salina* 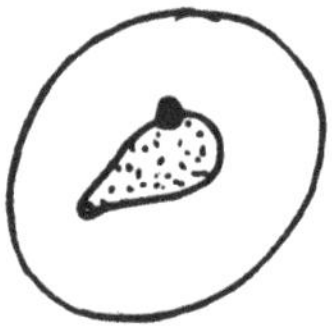	Division : Thallophyta Class : Chlorophyceae Genus : *Chlorella* Species : *salina*	• Cell is very small with a cell wall. • Cytoplasm is concentrated at the centre of the cell. • Nucleus present at one end of cytoplasm.
102.	*Chlorella marina* 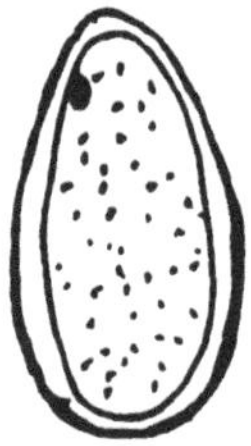	Division : Thallophyta Class : Chlorophyceae Genus : *Chlorella* Species : *marina*	• Presence of a cell wall. • Body of cell is ovate and green. • This is non-flagellate and mucilagenous.

4.2 Common zooplankton of northwestern Bay of Bengal

The zooplankton are important planktonic members of the aquatic medium, which are either herbivores, grazing on the phytoplankton, or carnivores feeding on other members of zooplankton. Many of the zooplankton have some ability to swim and can dart rapidly over short distances in pursuit of prey or to flee predators. They are also transported by the currents.

Some zooplankton migrate towards the sea surface each night and return to the depths each day, either in an attempt to maintain their light level or in response to the movement of their food resource. Foraminiferans and radiolarians are single-celled microscopic members of marine and estuarine zooplankton.

All of the major and most minor animal phyla are represented in the zooplankton community. A brief investigation of coastal zooplankton offers at the same time a review of general zoology, a survey of the invertebrates and an introduction to invertebrate embryology. The vertebrates are represented by fish eggs and young fry.

Planktonic animals are either meroplanktonic or holoplanktonic. Meroplankton spend only a portion of their life-cycles, usually the larval period, as the plankton. Examples of meroplankton include crab larvae, which settle after metamorphosis to become part of the **benthic** adult population, and fish larvae, which can settle out of the plankton to become part of the **nektonic** adult community. Many meroplankton spawn eggs in the planktonic form where they are fertilized by a male gamete. This zygote then develops through the blastula and gastrula stages to a larval form. Often larval forms are dramatically different in appearance from the familiar adults. After passing few weeks/months as the plankton, **metamorphosis** occurs and a young adult, more like the parental form, emerges. Holoplankton, on the other hand, spend their entire life-cycles in the planktonic form. Examples of holoplankton include copepods and chaetognaths.

In the northwestern Bay of Bengal, adjacent to deltaic Sundarbans, the zooplankton community comprises a heterogenous assemblage of animals covering many taxonomic groups including copepods, mysids, lucifers, gammarid, amphipods, cladocerae, ostracods, cumacea, hydromedusae, ctenophores and chaetognaths among holoplankters and polychaete larvae, molluscan and echinoderm larvae, crustacean larvae and fish eggs and larvae among meroplankters. Among the zooplankton, calanoid copepods constitute the major bulk, followed by cyclopoids and harpecticoids (Chaudhuri and Choudhury, 1994). In general, the higher abundance of zooplankton is encountered during the premonsoon periods. The main cause of the high species diversity index of zooplankton during the premonsoon months may be attributed to the coincidence of the breeding period of most of the coastal and estuarine fishes in the months of late February and March (Mitra, 2000).

Some common zooplankton found in the marine environment of coastal West Bengal are listed in Table 4.2.1.

TABLE 4.2.1 : Common zooplankton around Northwestern Bay of Bengal coast

S.No.	Name	Systematic position	Salient features
1.	*Globigerina triloculinoides* 200 μm	Phylum-Protozoa Class-Sarcodina Subclass-Rhizopoda Order-Forminifera Family-Orbulinidae Subfamily-Globigerininae Genus- *Globigerina* Species- *triloculinoides*	• Composed of subglobular chambers arranged in two whorls. • Wall is calcareous and coarsely perforated. • Sutures are distinct, depressed and umblical with a prominent lip.
2.	*Globigerina parva* 150 μm	Phylum-Protozoa Class-Sarcodina Subclass-Rhizopoda Order-Forminifera Family-Orbulinidae Subfamily-Globigerininae Genus- *Globigerina* Species-*parva*	• Test is small and highly lobate. • Wall is calcareous, finely perforated. • Presence of 12 chambers which are spherical and arranged in 2½ whorls.

S.No.	Name	Systematic position	Salient features
3.	*Globigerina opima*  200μm	Phylum-Protozoa Class-Sarcodina Subclass-Rhizopoda Order-Forminifera Family-Orbulinidae Subfamily-Globigerininae Genus- *Globigerina* Species-*opima*	• Equatorial periphery is slightly lobulate and periphery rounded. • Spherical chambers are arranged in $2\frac{1}{2}$ whorls of which last consists of $5\frac{1}{2}$ chambers. • Sutures are radially depressed and aperture is interomarginal and extraumbilical.
4.	*Acanthometron* sp. 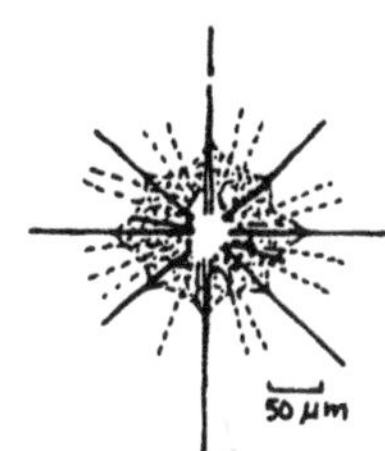 50 μm	Phylum-Protozoa Class-Sarcodina Subclass-Actinopoda Order-Radiolaria Genus-*Acanthometron* sp.	• Protoplasmic strands project in all directions as thin, long straight filaments. • Possesses beautiful and siliceous skeleton. • Central capsule is perforated and horny.
5.	*Tintinnopsis beroidea* 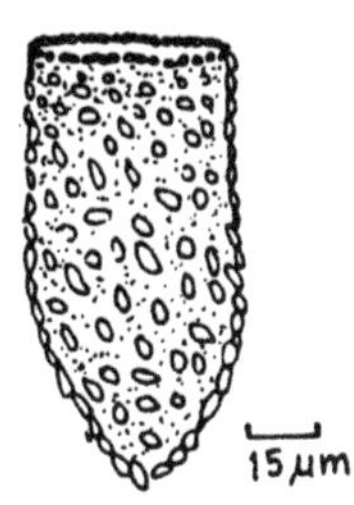 15 μm	Phylum-Protozoa Class-Polyhymenophora Subclass-Spirotrichia Order-Oligotrichida Suborder-Tintinnida Family-Codonellidae Genus-*Tintinnopsis* Species-*beroidea*	• Usually bullet shaped with slightly pointed end. • Encrustations heavy. • Aboral part of lorica remains spherical, while oral portion sometimes starts narrowing.
6.	*Tintinnopsis tobulosa*  25 μm	Phylum-Protozoa Class-Polyhymenophora Subclass-Spirotrichia Order-Oligotrichida Suborder-Tintinnida Family-Codonellidae Genus-*Tintinnopsis* Species-*tobulosa*	• Lorica possesses two regions – bowl and column. • Oral flare absent. • Bowl and column uniformly agglomerated.

S.No.	Name	Systematic position	Salient features
7.	*Brachionus angularis*  200 µm	Phylum-Rotifera Class-Monogononta Order-Ploima Family-Brachionidae Subfamily-Brachioninae Genus-*Brachionus* Species-*angularis*	• Lorica circular in shape. • Presence of four anterior spines, from which two marginal and two medium spines come out. • Presence of deep U-shaped sinus franked by dorsal median spines.
8.	*Brachionus calyciflorus* 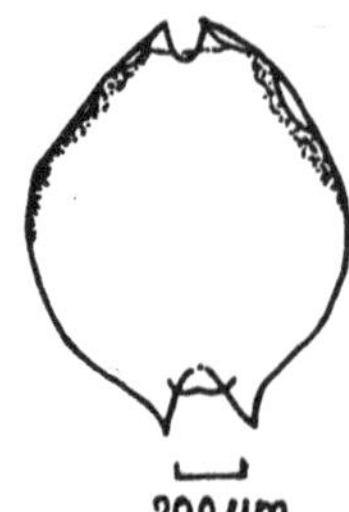 200 µm	Phylum-Rotifera Class-Monogononta Order-Ploima Family-Brachionidae Subfamily-Brachioninae Genus-*Brachionus* Species-*calyciflorus*	• Length and breadth of lorica, more or less the same. • Four occipital spines are present. • V-shaped sinus with median spines longer than lateral spines.
9.	*Nannocalanus minor* 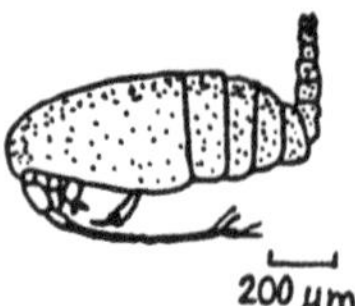 200 µm	Phylum-Rotifera Class-Monogononta Order-Ploima Suborder-Calanoida Family-Calanidae Genus-*Nannocalanus* Species-*minor*	• First antennae reaches upto caudal rami by about half of the body length. • Left leg distinctly longer than the right leg. • External marginal spines are greatly enlarged on left exopodite.
10.	*Acrocalanus gracilis* 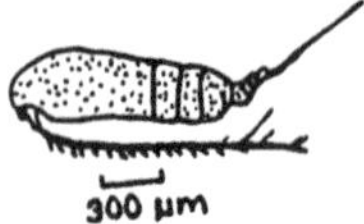 300 µm	Phylum-Rotifera Class-Monogononta Order-Ploima Suborder-Calanoida Family-Paracalanidae Genus-*Acrocalanus* Species-*gracilis*	• Body is parallel sided, almost 3 times as long as broad. • Cephalosome is evenly rounded. • Antennae longer than the body and having two long hairs at their ends.

S.No.	Name	Systematic position	Salient features
11.	*Paracalanus parvus* 	Phylum-Rotifera Class-Monogononta Order-Ploima Suborder-Calanoida Family-Paracalanidae Genus-*Paracalanus* Species-*parvus*	• Urosome 4-segmented in male and 5-segmented in females. • 5th leg symmetrical in male and asymmetrical in female. • Left foot much longer; bulb-like eminence on cephalosome.
12.	*Acartia spinicauda* 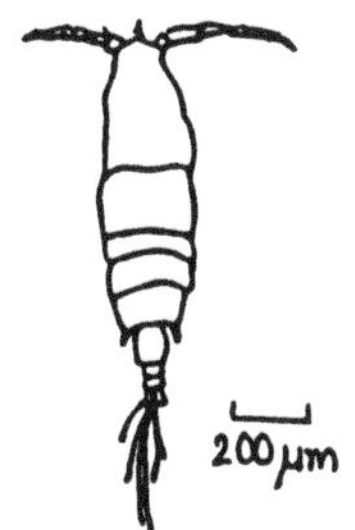	Phylum-Rotifera Class-Monogononta Order-Ploima Suborder-Calanoida Family-Acartiidae Genus-*Acartia* Species-*spinicauda*	• Spines are present at the corners of the metasome. • Terminal claw of the 5th leg straight and scarcely widened at base. • Males have spines at 3rd urosome segment which overreach the 4th segment.
13.	*Acartia erythraea* 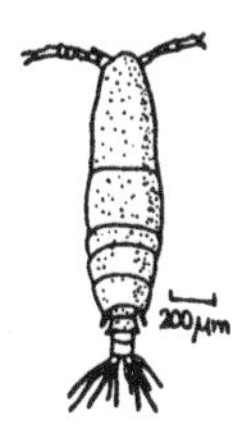	Phylum-Rotifera Class-Monogononta Order-Ploima Suborder-Calanoida Family-Acartiidae Genus-*Acartia* Species-*erythraea*	• Terminal claw of 5th leg is slightly curved and smooth. • 5th leg has a thickened terminal claw. • Presence of 2 pairs of prominent spines in 2nd urosome.
14.	*Acartia danae* 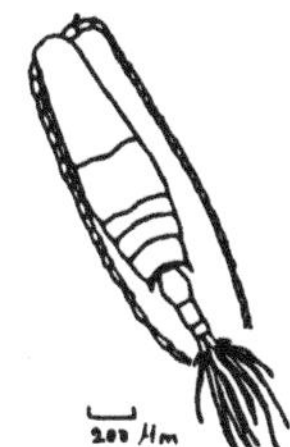	Phylum-Rotifera Class-Monogononta Order-Ploima Suborder-Calanoida Family-Acartiidae Genus-*Acartia* Species-*danae*	• 5th leg of female are straight. • Presence of crowd of small teeth near the tip. • Length of the species is 1.0 - 1.2 mm.

S.No.	Name	Systematic position	Salient features
15.	*Oithona similis*	Phylum-Rotifera Class-Monogononta Order-Ploima Suborder-Cyclopoida Family-Oithonidae Genus-*Oithona* Species-*similis*	• 1st antennae reaches upto end of the last metasome. • Body highly pelucid. • First segment bears a small semicircular process.
16.	*Leuckartiara sp.*	Phylum-Cnidaria Class-Hydrozoa Order-Anthomedusae Genus-*Leuckartiara* sp.	• Medusae may reach a height of 15 mm and possess highly folded margins. • Presence of about 20 marginal tentacles. • Gonads are horse-shoe shaped with folds directed outwards.
17.	*Aequorea* sp.	Phylum-Cnidaria Class-Hydrozoa Order-Anthomedusae Genus-*Aequorea* sp.	• Presence of saucer-shaped medusae of 20 cm diameter. • Presence of thick jelly in the bell. • Numerous radial canals are present attached to the gonads.
18.	*Cosmetira* sp.	Phylum-Cnidaria Class-Hydrozoa Order-Anthomedusae Genus-*Cosmetira* sp.	• Presence of numerous marginal tentacles (about 100) with swollen bases. • Presence of light marginal vesicles without ocelli. • Gonads present on the radial canals are long.

S.No.	Name	Systematic position	Salient features
19.	*Obelia medusa*	Phylum-Cnidaria Class-Hydrozoa Order-Anthomedusae Genus-*Obelia* Species-*medusa*	• Medusae are saucer-shaped with a very flat bell. • There are 4 radial canals and one ring canal from which 24 solid marginal tentacles originate. • Body is transparent, medusae luminescent with 8 abradial marginal vesicles.
20.	*Phialidium* sp.	Phylum-Cnidaria Class-Hydrozoa Order-Anthomedusae Genus-*Phialidium* sp.	• Diameter of the umbrella may reach 20 mm but is normally between 5-10 mm. • Presence of 32 hollow marginal tentacles and upto 3 marginal vesicles found between each pair of tentacles. • Presence of 4 radial canals.
21.	*Physalia physalis*	Phylum-Cnidaria Class-Hydrozoa Order-Siphonophora Sub Order-Physophorida Family-Physaliidae Genus-*Physalia* Species-*physalis*	• Presence of enormous coloured float formed simply as a hollow pocket by infolding of outer layer. • At regular intervals, the whole body twists over to wet itself first on one side and then on the other. • Presence of hydrophyllia or bracts which protect the gonophores and gastrozooids.

The figures in the Name column show drawings with scale bars labelled "2 mm" (row 19), "5 mm" (row 20), and "10 cm" (row 21).

S.No.	Name	Systematic position	Salient features
22.	*Porpita porpita* 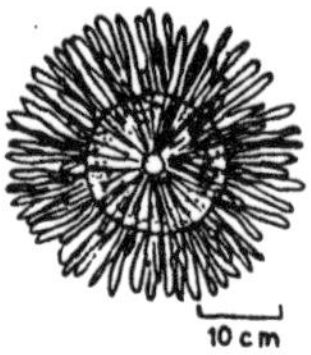 	Phylum-Cnidaria Class-Hydrozoa Order-Athecata Suborder-Capitata Family-Velellidae Genus-*Porpita* Species-*porpita*	• It is a free-floating modified hydroid polyp upto 8 cm in diameter. • Numerous clavate tentacles bearing powerful stinging cells hang from the rim of horny central disc-shaped float which support the animal on the water surface. • Body tissues contain symbiotic algae.
23.	*Velella velella* 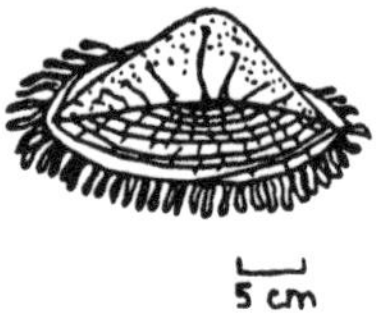 	Phylum-Cnidaria Class-Hydrozoa Order-Athecata Suborder-Capitata Family-Velellidae Genus-*Velella* Species-*velella*	• It is a complex colony made up of a large number of polyps crowding the under surface of an oval-shaped float. • Presence of central mouth with large number of smaller polyps. • Deep blue in colour with a transluscent sail protruding above the water surface.
24.	*Aurelia aurita* 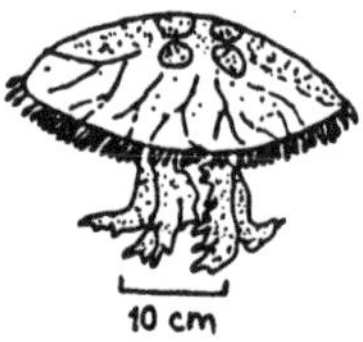 	Phylum-Cnidaria Class-Scyphozoa Order-Semaeostomeae Family-Ulmaridae Genus-*Aurelia* Species-*aurita*	• Bell is saucer-shaped and about 50 cm in diameter. • Can be recognised by its bright purple to pale lilac-coloured ovaries or testis. • Presence of numerous fine short tentacles and 8 sense organs around the edge of the umbrella.

S.No.	Name	Systematic position	Salient features
25.	*Rhizostoma sp.* 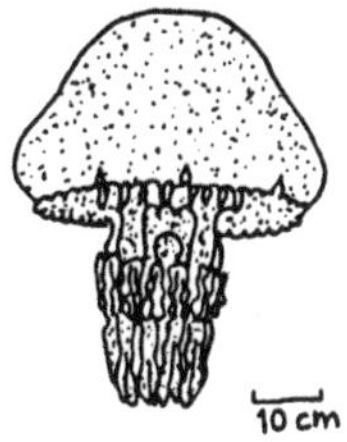	Phylum-Cnidaria Class-Scyphozoa Order-Rhizostomeae Genus-*Rhizostoma* sp.	• Presence of dome-shaped bell upto 1m in diameter. • Deep purplish blue in colour with margins lobed and devoid of tentacles. • Presence of multi-ciliated mouth that act as suckers for nourishment.
26.	*Beroe* sp. 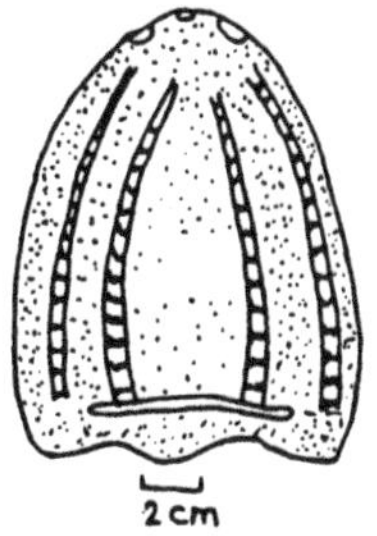	Phylum-Cnidaria Class-Scyphozoa Order-Rhizostomeae Family-Beroidae Genus-*Beroe* sp.	• Presence of 8 rows of beating plates down the two sides. • It is about 15 cm in length and shaped like vegetable marrow or thimble. • Mouth opens into a large space that fills the entire mass of the animal body.
27.	*Tomopteris* sp. 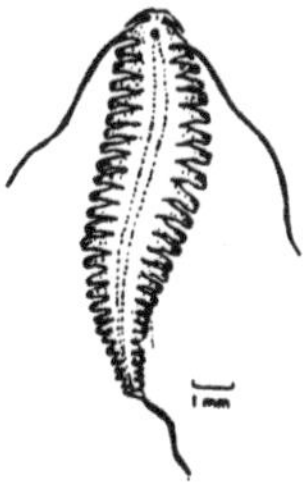	Phylum-Annelida Class-Polychaeta Order-Phyllodocida Family-Tomopteridae Genus-*Tomopteris* sp.	• Body is transparent and parapodia are flat and transluscent. • Parapodia are paddle-shaped, biramous and lack setae. • Presence of 2 long tentacles, one on each side of the head.
28.	*Sagitta enflata* 	Phylum-Chaetognatha Class-Sagittoidea Genus-*Sagitta* Species-*enflata*	• Commonly known "arrow-worm" because of its long, straight, slender body. • Body is divisible into short head, long trunk and a short tail piece. • Mouth is with stout curved bristles and body with two pairs of horizontal side fins and a tail fin.

S.No.	Name	Systematic position	Salient features
29.	*Sagitta maxima* 	Phylum-Chaetognatha Class-Sagittoidea Genus-*Sagitta* Species-*maxima*	• Head is wide and oval with seminal vesicles distinct. • Posterior fin merges with the anterior one. • Characterised by the presence of an anterior pair of lateral fins which are small and rounded.
30.	*Sagitta lyra* 	Phylum-Chaetognatha Class-Sagittoidea Genus-*Sagitta* Species-*lyra*	• Body is transluscent and somewhat flabby with greater width at the middle. • Head is large and wider than long with eyes oval and pigmented; neck is well-defined. • Antero-lateral fins are longer than the posterior fins and tail fin is lobate posteriorly.
31.	*Sagitta hispida* 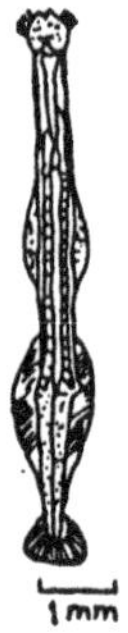	Phylum-Chaetognatha Class-Sagittoidea Genus-*Sagitta* Species-*hispida*	• Body is stout, rigid and opaque. • Head is wider than the body and there is no distinct neck. • Anterior fins are shorter and narrower than the posterior fins.

S.No.	Name	Systematic position	Salient features
32.	*Sagitta robusta*	Phylum-Chaetognatha Class-Sagittoidea Genus-*Sagitta* Species-*robusta*	• Head not broad but body opaque due to presence of strong longitudinal muscles. • Anterior and posterior lateral fins are about equal in length and fully rayed. • Eyes are very large and eye pigment is concentrated in an eclipse.
33.	*Pleurobrachia pileus*	Phylum-Ctenophora Class-Tentaculata Order-Cydippida Family-Pleurobrachiidae Genus- *Pleurobrachia* Species-*pileus*	• Popularly known as "comb jelly" or "sea gooseberry". • Beautiful, delicate, bio-luminating planktonic organism. • Presence of eight rows of comb plates and a balancing organ (statocyst).
34.	*Podon* sp.	Phylum-Arthropoda Class-Crustacea Subclass-Branchiopoda Order-Cladocera Family-Polyphenidae Genus-*Podon* sp.	• Head demarcated from the body by a deep transverse groove. • Bivalved carapace forms a distinct dorsal chamber which acts as brood pouch. • Presence of biramous setose antennae behind the prominent compound eye.

S.No.	Name	Systematic position	Salient features
35.	*Gigacuma halei* 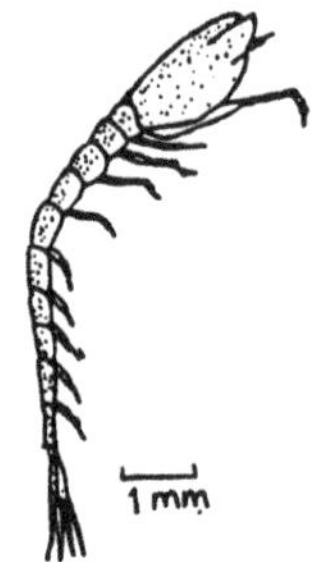	Phylum-Arthropoda Class-Crustacea Subclass-Branchiopoda Order-Cumacea Genus-*Gigacuma* Species-*halei*	• Presence of large swollen cephalothorax and long slender abdomen ending in style-like caudal rami. • Possesses carapace which covers 3 or 4 thoracic segments and anteriorly first thoracic limb. • First 3 pairs of thoracic limbs are modified to maxillipeds and uropods are slender.
36.	*Euphausia diomediae* 	Phylum-Arthropoda Class-Crustacea Subclass-Branchiopoda Order-Euphausiacea Genus-*Euphausia* Species-*diomediae*	• Possesses external gills attached to endopods of biramous thoracic limbs. • Presence of prominent pair of spines on each laterofrontal border and smaller one on each lateral margin of the carapace. • They are of about 40 mm in length.
37.	*Lucifer hanseni* 	Phylum-Arthropoda Class-Crustacea Subclass-Branchiopoda Order-Decapoda Suborder-Dendrobranchiata Genus-*Lucifer* Species-*hanseni*	• Body is laterally compressed. • It is about 1 cm in length. • The 5th pereiopod is absent or vestigial and luminescent cells are present in the telson.

S.No.	Name	Systematic position	Salient features
38.	*Macrosetella gracilis* 	Phylum-Arthropoda Class-Crustacea Order-Harpacticoida Genus- *Macrosetella* Species- *gracilis*	• Body is fusiform with length about 1.2-1.3 mm (males) and 1.4 – 1.5 mm (females). • Caudal setae are very long. • Caudal rami are slender, cylindrical and 4 times as long as broad.

Larval forms

S.No.	Name	Systematic position	Salient features
39.	Trochophore larvae of polychaete 	Phylum-Annelida Class-Polychaeta Specimen- Trochophore larvae of polychaete	• Almost spherical body with little tuft of long cilia and sensory cells at the upper pole. • Mouth is located equatorially. • A main ciliated girdle runs around the sphere just above the equator.
40.	Spirorbis larvae 	Phylum-Annelida Class-Polychaeta Specimen-Spirorbis larvae	• Larva exhibits bilateral symmetry and can be divided into head, thorax and abdomen. • The prototroch consists of a complete circle of cilia. • Branchial rudiments are present under the dorsolateral sides of the head.

S.No.	Name	Systematic position	Salient features
41.	Sabellarid larvae 	Phylum-Annelida Class-Polychaeta Specimen-Sabellarid larvae	• Characterised by the presence of 4 red eye-spots and a pair of posteriorly projecting tentacles. • Presence of 2 bundles of barbed provisional setae. • Presence of a dorsal hump posterior to the eye-spot and two dorsal longitudinal rows of highly clustered cilia.
42.	Copepod nauplii 	Phylum-Arthropoda Class-Crustacea Order-Copepoda Specimen-Copepod nauplii	• Body kite shaped with pointed end and single distinct eye on the front side. • Presence of three pair of limbs. • First pair of limb is uniramous and others biramous.
43.	Nauplii of *Penaeus indicus*	Phylum-Arthropoda Class-Crustacea Order-Decapoda Suborder-Macrura Specimen-Nauplii of *Penaeus indicus*	• Ocellus present at the anterior median region of the body. • Presence of a pair of dorsally curved caudal setae at the posterior end of the body. • First pair of appendage is uniramous, 2nd and 3rd pair is biramous, with the third pair shorter than the other appendage.
44.	Nauplii of Balanus 	Phylum-Arthropoda Class-Crustacea Order-Decapoda Suborder-Macrura Specimen-Nauplii of Balanus	• Body is triangular. • Presence of a pair of posterior spines. • Tip of the rostrum is truncated.

S.No.	Name	Systematic position	Salient features
45.	Cypris of Balanus 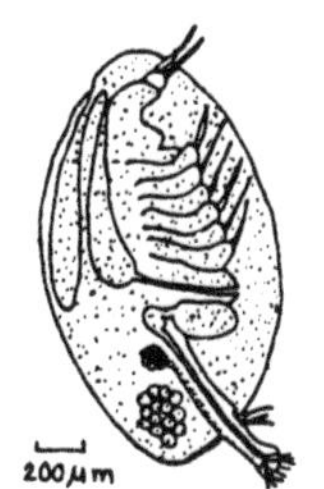	Phylum - Arthropoda Class - Crustacea Order-Decapoda Suborder-Macrura Specimen-Cypris of Balanus	• Presence of a bi-valve shell. • Presence of a compound eye and 6 pairs of thoracic appendages. • Consists of a short abdomen.
46.	Larva of *Squilla hieroglyphica* 	Phylum-Arthropoda Class-Crustacea Order-Stomatopoda Specimen-Larva of *Squilla hieroglyphica*	• Carapace large and broad. • Presence of 3 spinules on lateral margin of the carapace — 1st near the base of anterolateral spine, 2nd at the junction between 5th and 6th thoracic somites and 3rd at the posterolateral corner. • Eye-stalk is ¾ as long as the cornea.
47.	Protozoea of *Penaeus indicus* 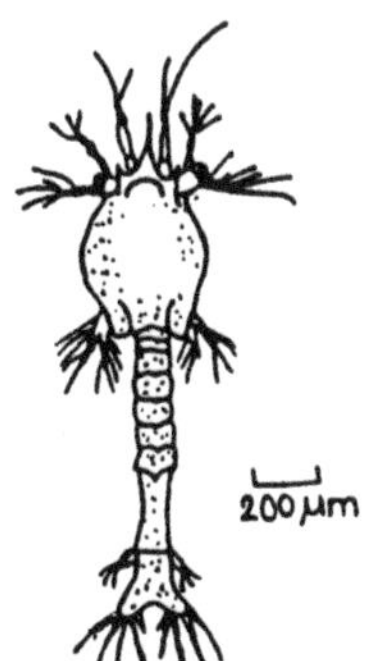	Phylum-Arthropoda Class-Crustacea Order-Decapoda Suborder-Macrura Specimen-Protozoea of *Penaeus indicus*	• Presence of well developed curved rostrum. • Presence of bifurcated supraorbital spines and stalked compound eyes. • Length is 1.5 mm and there are no frontal organs.

S.No.	Name	Systematic position	Salient features
48.	Mysis of *Penaeus indicus*	Phylum-Arthropoda Class-Crustacea Order-Decapoda Suborder-Macrura Specimen-Mysis of *Penaeus indicus*	• Presence of spine on the scaphocerite and an unsegmented pleopod bud. • Cleft of telson extends to the origin of the penultimate pair of lateral telsonic setae. • Length of the organism is 3.5 mm.
49.	Zoea of mud crab, *Scylla serrata*	Phylum-Arthropoda Class-Crustacea Order-Cirripedia Suborder-Brachiura Specimen- Zoea of mud crab, *Scylla serrata*	• Presence of laterally compressed carapace. • Characterised by one dorsal, one rostral and 2 lateral spines, all of which are elongated. • Eyes sessile, abdomen 5-segmented and telson is furcated.
50.	Megalopa of mud crab, *Scylla serrata*	Phylum-Arthropoda Class-Crustacea Order-Cirripedia Suborder-Brachiura Specimen- Megalopa of mud crab, *Scylla serrata*	• All appendages are well developed. • Abdomen with 5 pairs of swimming pleopods. • Carapace depressed and a pair of chelipeds also develops.
51.	Veligers of *Saccostrea cucullata*	Phylum-Mollusca Class-Bivalvia Specimen-Veligers of *Saccostrea cucullata*	• It is a D-shell larval stage with semi-transparent velum and is about 70 μm size. • Oval shape at early "umbo stage" when measures 100 μm. • Presence of irregular eye-spot at "eyed stage" when it measures 300 μm.

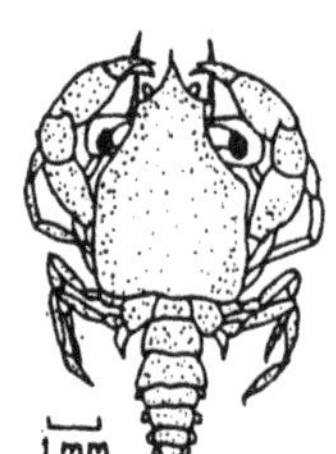

S.No.	Name	Systematic position	Salient features
52.	Veliger of *Littorina littoraeia* 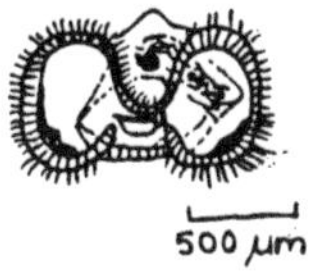	Phylum-Mollusca Class-Bivalvia Specimen-Veligers of *Littorina littoraeia*	• Larvae is 1½ whorled shape. • Characterised by 2 dark pigmented particles. • It is pale yellowish and has no sculpturing.
53.	Bipinnaria larva of starfish 	Phylum-Echinodermata Class-Asteroidea Specimen- Bipinnaria larva of starfish	• Presence of a wavy band of cilia at the equatorial girdle. • This band runs down on each side and loops round the front above and below where the mouth begins to form. • Before the mouth opens, gut is developed with a tiny stomach and intestinal opening at the lower end of larva.
54.	Brachiolaria larva 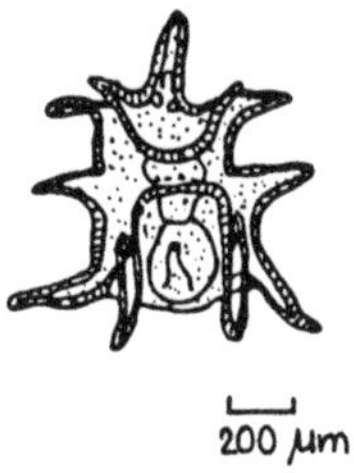	Phylum-Echinodermata Class-Asteroidea Specimen- Brachiolaria larva	• Composed of 3 special appendages in the preoral region called brachiolarian arms. • Presence of adhesive papillae on the arms except for the medio-dorsal one. • The arms are unciliated except the medio-dorsal one that bears circum oral ciliary band.

S.No.	Name	Systematic position	Salient features
55.	Auricularia larva of sea cucumber	Phylum-Echinodermata Class-Holothuroidea Specimen- Auricularia larva of sea cucumber	• The foldings of the ciliated band is very long. • The circumoral ciliary band is drawn out into paired pre-oral lobes anteriorly and paired anal lobes posteriorly. • Circumoral ciliary field is continuous.
56.	Brachyuran zoea	Phylum-Rotifera Class-Monogononta Order-Ploima Suborder-Brachyura Family-Portunidae Specimen-Brachyuran zoea	• Single dorsal and rostral spine elongated. • A pair of lateral spines present, one pair pointed downwards and other pointed upwards and forwards. • Ends of spines are flattened like spear head.
57.	Fish egg and fish larvae	Phylum-Chordata Class-Pisces Specimen- Fish egg and fish larvae	• Generally spherical and contain yolk & oil globules. • Presence of prominent eyes, mouth and tail fin. • Pigmented, head, body and pectoral region.

Remark: Defining a species from the taxonomic point of view varies from author to author and also amongst the different working groups. This is for the information of the readers that this classification is a traditional one and has been followed from two important books namely "Phytoplankton of the Indian seas" and "A manual of marine zooplankton."

Suggested references

1. Banerjee, Kakoli; Mitra, Abhijit; Bhattacharyya, D. P. and Amalesh Choudhury. (2002). Role of nutrients on phytoplankton diversity in the North east coast of the Bay of Bengal. *In: Ecology and Ethology of Aquatic Biota* (Editor Arvid Kumar, IJEE), Chapter – 6. pp. 102-109.

2. Chaudhuri, A. B and Choudhury, A (1994). Mangroves of the Sundarbans, India. IUCN.

3. Mitra, A (2000). The Northwest coast of the Bay of Bengal and deltaic Sundarbans. In: Seas at the Millennium – An Environmental Evaluation, Chapter 62 (Editor: Charles Sheppard, University of Warwick, Coventry, UK), Elsevier Science, 143 – 157.

4. Mitra, Abhijit; Banerjee, Kakoli; Choudhury, Amalesh and D. P. Bhattacharyya. (2002). Taxonomy of coastal phytoplankton inhabiting polluted waters. In: Ecology of Polluted Waters (Editor Arvind Kumar, IJEE), Chapter 78.

Internet references

- *http://www.iucn.org/themes/wetlands/sundarbans1.htm*
- *http://www.wcmc.org.uk/igcmc/s_sheet/worldh/sundarba.html*

Chapter 5
ROLE OF NATURAL PROCESSES ON PLANKTON COMMUNITY

Contents

5.1 Effect of Upwelling and El Nino on marine plankton

Upwelling is the phenomenon of transference of water mass from the bottom part of the ocean to the surface layer caused by the difference in density of the water layer triggered mainly by variation of temperature at different depths. Vertical circulation driven by surface changes in temperature and salinity is known as **thermohaline circulation**. If a surface process forms more-dense water on top of less-dense water, the water column becomes unstable and as the denser water sinks or **downwells**, the less dense water rises or **upwells**. This is infact known as **overturning** and the process brings lot of nutrients from the bottom layer to the surface layer of the oceans. The planktonic diversity in such zone is extremely rich owing to the presence of high concentrations of nitrate, phosphate and silicate in the aquatic phase. In the Atlantic Ocean, waters of different densities are formed at the surface at different latitudes. They sink and flow north or south at differing depths. Arctic Ocean layers are controlled by the seasonal freezing and thawing of the sea ice. The water mass of the Pacific Ocean moves in a very sluggish fashion. The water layers of the Indian Ocean are also less distinct than those of the Atlantic, with no water that corresponds to that formed in the northern latitudes.

Seasonal upwellings and downwellings are very common in coastal zones due to change of wind patterns and alternating flow of water onshore and offshore due to the Ekman transport. The areas of coastal upwelling exhibit high phytopigment concentrations. The reason behind this is that due to coastal upwelling considerable concentrations of nutrient are provided by vertical transport of subsurface water into the euphotic layer, and optimal light conditions for phytoplankton growth are maintained in the stabilized horizontal divergent flow of the surface layer. Optimal nutrient and light conditions are maintained for long periods of time each year, so that amount and pattern of biological productivity are distinct from other regions of the ocean. Researchers who have worked on this particular topic, such as Margalef (1978 a, b), Vinogradov and Shushkina (1978), Walsh (1977) and Boje and Tomczak (1978), agree that upwelling zones have special types of identity from other zones of the marine ecosystem. The major upwelling zones of the world are restricted to the west coasts of North and South America, the west coast of Africa and along the western side of the Indian Ocean. These are the major upwelling areas on the sheltered sides of landmasses in the trade wind belts and in areas of abundant sunlight. In these nutrient rich zones, the rich population of phytoplankton form the base of biological production, which is also reflected through abundant fish catches in this zone.

Observational and modern studies have demonstrated how the coastal upwelling circulation is exploited by zooplankton (Peterson *et al.*, 1979; Brockmann, 1979) and fish (Mathisen, 1980) to maintain themselves in the ecosystem. Of the zooplankton and nekton that characterise upwelling, it has been observed that euphausiids are specially adapted to coastal upwelling ecosystems. At or just beyond the shelf break in all major coastal upwelling regions, large populations of euphausiids are present (Thiriot, 1978). Euphausiids consume as much of the productivity as anchoveta and because of their habitat in the shelf-break region, they are the major link through which shelf productivity is transferred to fish such as jack mackeral and hake. Thus, a direct relationship of planktonic richness with upwelling phenomenon can be drawn. The downwelling centers of the large ocean gyres, on the other hand, are restricted to those areas where surface convergences occur. These areas are characterised by low nutrient availability that depresses primary production. It has been documented that on an average, upwelling areas are four times more productive than coastal areas and five times more productive than the open ocean for the same units of area and time (Table 5.1.1).

TABLE 5.1.1

Area	Primary production (gC/m²/yr)	World ocean area (km²)	(%)	Total primary production (metric tone of carbon/year)
Upwellings	640	0.36 X 10^6	0.1	0.23 X 10^9
Coasts	160	54 X 10^6	15.0	8.6 X 10^9
Open oceans	130	307 X 10^6	85.0	39.9 X 10^9

Source: Data from S. Smith and J. Hollibaugh, "Coastal Metabolism and the Oceanic Organic Carbon Balance" in Review of Geophysics 31 (1): 75-89, 1993.

The table 5.1.1 reveals the fact that the rate of primary production in the marine ecosystem varies in the order: upwelling zone > coastal zone > open ocean. Interested students may compare this trend with the terrestrial ecosystem where the trend of primary production rate is of the sequence: heavily cultivated land > pasture land > desert area. Thus the open ocean is comparable to desert ecosystem on land. However, because of the large area of the open oceans (307 X 10^6 km²), the total generation of organic carbon by phytoplankton community is about 39.9 X 10^9 metric tons of carbon / year, which is approximately 174 times higher than that produced in the upwelling zones.

El Nino is a severe atmospheric and oceanic disturbance in the Pacific Ocean that occurs every 7-14 years. The phenomenon is called El Nino meaning "the Christ Child" because it usually appears during the Christmas season. The exact cause of El Nino is still not clear, but certain processes have been identified with its appearance. When atmospheric pressure increases over Eastern Island in the eastern Pacific, decreases over Indonesia in the western Pacific and then reverses, it is known as **Southern Ocean Oscillations**. It causes West Pacific surface water to surge across the Pacific and raise surface water temperatures along South America. Because of the relationship between El Nino and Southern Oscillation, the two events together are known as ENSO. During El Nino years, the marine and estuarine ecosystems

experience the following important events (http://
www.pmer.noaa.gov/toga-tao/er-mno-report.html):

(a) the thermocline along the equator flattens out, rising in the
     west and plunging several hundred feet below the surface
     in the east – deep enough so that the coastal upwelling is
     no longer able to tap the cold, nutrient rich waters from
     beneath it.

(b) Equatorial upwelling decreases reducing the supply of
     nutrients to the euphotic zone.

(c) Sea level flattens out, dropping in the west and rising in
     the east. Surface water surges eastwards along the equator.
     These events cause warm surface waters to flow from the
     Central Pacific towards the eastern Pacific, supressing the
     cold nutrient rich upwelling of the Homboldt current off
     the coast of South America. This disturbance leads to a fall
     in the planktonic diversity due to which the marine food
     chain is greatly affected. A report published by Randolph
     E. Schmid on 10th December, 1999 (http://www.djc.com/
     news/enviro/11001529.html) under the article "El Nino
     causes plankton boom-bust cycle and affects carbon-di-
     oxide" states that the periodic El Nino warming of the
     Pacific Ocean not only reduces the amount of carbon-di-
     oxide in the air, but also causes a boom and bust cycle for
     the tiny free floating marine plants.

A team of researchers led by Francisco Chavez of the Monterey
Bay Aquarium Research Institute found that the 1997-98 El Nino
and the subsequent ocean cooling called the La Nina, had a roller-
coster effect on the oceanic food chain across the vast swath of
Pacific. According to this research team, during El Nino warm
episode, the normal upwelling of cold deep ocean water is blocked,
due to which the availability of nutrients to the planktonic
community is obstructed. This sharply decreases the biomass and
diversity of phytoplankton in the area. In the final stage, this
episode leads to a disturbance of the marine food chain due to
which fishes become less abundant.

5.2. Impact of monsoon on marine plankton

Without the atmosphere, there would have been no life at all
on the earth. The blanket of air surrounding the earth protects

The highest diversity of phytoplankton during premonsoon in these case studies is due to optimum transparency and salinity (Banerjee *et al.*, 2002). Although no studies have been conducted to know the pattern of zooplankton oscillation with reference to phytoplankton community and monsoonal impact in north western Bay of Bengal, but abundance and distribution pattern of zooplankton exhibit a significant positive correlation and covariance with that of nutrients and phytoplankton (map.seafdec.org/fishoceano/zoo.htm). Few works carried out in the coastal waters of West Bengal, a maritime state in the Northeast part of India, reveal high diversity index of ichthyoplankton during postmonsoon and premonsoon (Sasmal *et al.*, 1998; Mitra *et al.*, 1999). Simultaneous analyses of phytoplankton and zooplankton standing crop and biomass may however, establish a correct picture on the inter-relationship between these two different types of planktonic community and the grazing pressure exerted by zooplankton on the phytoplankton community.

Suggested references

1. Boje, R. and Tomczak, M. (1978). Ecosystem analysis and the definition of boundaries in upwelling regions. *In "Upwelling Ecosystems"* (R. Boje and M. Tomczak. eds) pp. 3 – 11. Springer – Verlag. New York.

2. Banerjee, Kakoli; Mitra, Abhijit; Bhattacharyya, D. P.; Das, Kiran Lal and Amalesh Choudhury. (2000). A preliminary study of phytoplankton diversity and water quality around Haldia port-cum-industrial complex. *National Seminar on Protection of the Environment – An Urgent Need*, organised by IPHE, Calcutta, pp. 15-18.

3. Banerjee, Kakoli; Mitra, Abhijit; Bhattacharyya, D. P. and Amalesh Choudhury. (2002). Role of nutrients on phytoplankton diversity in the North east coast of the Bay of Bengal. *In: Ecology and Ethology of Aquatic Biota* (Editor Arvind Kumar, IJEE), Chapter – 6. pp. 102-109.

4. Brockmann, C. (1979). A numerical upwelling model and its application to a biological problem. *Meeresforsch.* 27, 137-146.

5. Margalef, R. (1978a). Life-forms of phytoplankton as survivals alternatives in an unstable environment. *Oceanologica Acta* 1, 493–509.

6. Margalef, R. (1978b). What is an upwelling ecosystem ? *In "Upwelling Ecosystems"* (R. Boje and M. Tomczak, eds) pp.12–14. Springer – Verlag. New York

7. Mathisen, O. A. (1980). Adaptation of the anchoveta (*Engraulis ringens* J.) to the Peruvian upwelling system. *In: Productivity of Upwelling Ecosystems* (R. T. Barber and M. E. Vinogradov, eds). Elsevier, Amsterdam.

8. Mitra, Abhijit; Banerjee, Kakoli; Choudhury, Amalesh and D. P. Bhattacharyya. (2002). Taxonomy of coastal phytoplankton inhabiting polluted waters. In: Ecology of Polluted Waters (Editor Arvind Kumar, IJEE), Chapter 78.

9. Mitra, A; Sasmal, S. K; Choudhury, A; Mitra, Shampa and D. P. Bhattacharyya. (1999). Ichthyoplankton richness around Sagar Island, West Bengal, India. *Journal of Inland Fisheries Society of India*, 31 (1), pp. 38-42.

10. Peterson, W. T., Miller, C. B. and Hutchinson, A. (1979). Zonation and maintenance of copepod populations in the Oregon upwelling zone. *Deep-Sea Res.* 26 A. 467-494.

11. Sasmal, S. K; Kaviraj, A; Choudhury, A and Mitra, A. (1998). Community structure of icthyoplankton in the coastal zone of West Bengal, India. *River Behaviour and Control*, Vol. 25. pp. 80-82.

12. Thiriot, A. (1978). Zooplankton communities in the West African upwelling area. *In: Upwelling Ecosystems* (R. Boje and M. Tomczak, eds) pp. 32-61. Springer-Verlag, New York.

13. Vinogradov, M. E., Shushkina, E. A. (1978). Some development patterns of plankton communities in the upwelling areas of the Pacific Ocean. *Mar. Biol.* 48. 357 – 366.

14. Walsh. J. J. (1977). A biological sketchbook for an eastern boundary current. *In "The Sea"* (E. D. Goldberg, I. N. McCave, J. J. O'Brien and J. H. Steele, eds). Vol. 6, pp. 923 – 968. *Interscience*, New York.

Internet references

- *http://www.djc.com/news/enviro/11001529.html*
- *http://www. map.seafdec.org/fishoceano/phyto.htm*
- *http://www. map.seafdec.org/fishoceano/zoo.htm*

Chapter 6
PLANKTON CULTURE

Contents

6.1. Phytoplankton culture

A culture may be defined as an artificial environment in which the algae grow. In theory, culture conditions should resemble the alga's natural environment as far as possible. The culture of phytoplankton is an important aspect of planktonology and has immense commercial value. This is because mass culture of these organisms plays a vital role in the nutrition of finfish and shellfish especially in the early part of their life cycle.

The phytoplankton culture involves different phases such as (i) selection of suitable vessels and their sterilization, (ii) formulation and preparation of suitable culture media, (iii) isolation of the desired species for inoculation, (iv) management of culture in the laboratory vessels or in large outdoor tanks under controlled conditions of temperature, light and aeration and (v) harvesting of the cells at the different phases of growth.

(i) Selection of suitable vessels and their sterilization

Vessels such as test tubes, petridishes, conical flasks (1 litre) round and flat bottomed flasks (2 litre) and glass carbuoys (25 litre) which are made of 'Pyrex' or 'Corning glass' are normally used for culturing phytoplankton. For mass production of phytoplankton, large cone shaped fibre glass cylinders (100 litre) provided with a bottom outlet for emptying the culture, outdoor circular tanks (5000 litre) with bottom slopes to a centre outlet and concrete tanks are used. Glasswares plugged with cotton provide enough aeration for the growing cells. However, for carbuoys, cylinders and out door circular tanks, vigorous aeration may be required.

The vessels used for phytoplankton culture are cleaned first with chlorate-sulphuric acid followed by repeated rinsings with distilled water and sterilization in a hot air oven or in an autoclave for a period of 30 minutes under a pressure of 15 lb/m². Carbuoys, cylinders or outdoor circular tanks are normally cleaned by rinsing with distilled water followed by steam cleaning.

(ii) Formulation and preparation of suitable culture media

A number of media are now-a-days used for culturing the different groups of phytoplankton. Some of them are discussed here.

- *Schreber's medium*

This solution has the following composition.

Filtered seawater	1 litre
Sodium nitrate	0.1 gm
Sodium acid phosphate	0.02 gm
Soil extract	50 cc

The soil extract is prepared by dissolving 1 kg of good fertile garden soil in 1 litre of distilled water and autoclaving it for about 30 minutes under a pressure of 15 lb/m². After a day or two, the supernatant soil extract is decanted into a flask and autoclaved again for 30 minutes. It is always better to use freshly prepared soil extract for the preparation of culture medium. This medium is invariably used for culturing almost all phytoplankton species.

- *Miquel's medium*

It is comprised of two solutions, viz. A and B. The former consists of 20.2 gm of potassium nitrate dissolved in 100 ml of distilled water and the later is made of sodium phosphate (4 gm), calcium chloride (4 gm), ferric chloride (2 gm), concentrated hydrochloric acid (2 ml) all dissolved in 100 ml of distilled water. The medium is prepared by adding 0.55 ml of solution A and 0.50 ml of solution B to 1 litre of filtered seawater. This medium is also used for culturing several species of phytoplankton.

(iii) Isolation of desired species for inoculation

The isolation of required species for inoculation in the culture medium can be done by any of the following methods :

(a) Pipette method

By the pipette method, larger species are pipetted out using micropipettes under a microscope and are washed by transferring them into a series of sterilized cavity slides each containing autoclaved medium.

(b) Centrifugation method

Here water samples containing larger species are centrifuged at 3000 rpm and the sedimented organisms are washed by transferring them into a series of sterilized petridishes or watch glasses containing the autoclaved medium.

(c) Agar plating method

Several culture methods have been formulated to culture and grow the phytoplankton. **Agar plating method** is a widely used technology in which the agar medium is prepared by adding 1.5% agar to 1 litre of filtered seawater and this solution is autoclaved for about 15 minutes under a pressure of 15 lb/cm^2. This agar medium is then poured aseptically into the sterilized petridishes and are left undisturbed for a period of 24 hours. Instead of petridishes, test tubes may also be used. The agar medium is added to one-third of the test tube which is plugged with cotton before autoclaving.

The isolated species are picked up by a platinum needle or platinum loop under the microscope and streaked on the surface of the agar plates. After inoculation, these petridishes are placed in an incubation chamber maintained with constant temperature and light intensity for about 15 days when the streaked individual cells multiply leading to the formation of colonies. The later are removed by the platinum loop and are transferred to the culture tubes containing the medium aseptically. Alternately, the isolated and washed phytoplankton cells may be directly inoculated to the sterilized test tubes containing the medium. Periodically, the tubes containing the culture may be transferred to sterilized culture flasks or larger glass carbuoys containing the desired medium with a view to obtain mass cultures.

(iv) Management of culture in laboratory

In theory, culture conditions should resemble the alga's natural environment as far as possible; in reality many significant differences exist, most of which are deliberately imposed. A culture has three distinct components (a) a culture medium contained in a suitable vessel (b) the algal cells growing in the medium and (c) air, to allow exchange of carbon-dioxide between medium and the atmosphere. Extensive measures must be taken to keep pure unialgal cultures chemically and biologically clean. Chemical contamination may have unquantifiable, often deleterious and therefore undesirable effects on algal growth. Biological contamination of pure algal cultures by other eukaryotes and prokaryotic organisms in most cases invalidates experimental work, and may lead to the extinction of the desired algal species in culture through out-competition or grazing. In practice, it is very difficult to obtain bacteria-free (axenic) cultures, and although measures should be taken to minimize bacterial numbers, a degree of bacterial contamination is often acceptable. For an entirely autotrophic alga, all that is needed for growth is light, carbon-dioxide, water, nutrients and trace elements. By means of photosynthesis the alga will be able to synthesize all the biochemical compounds necessary for growth. Only a minority of alga seem, however, to be entirely autotrophic, many are unable to synthesize certain biochemical compounds (certain vitamins, for example) and will require these to be present in the medium. This condition is known as **auxotrophy**. Based on their growth characteristics, two kinds of cultures can be defined. In **limited volume (batch) cultures**, resources are finite. When the resources in the culture present in the culture medium are abundant, growth occurs according to the sigmoid curve, but once the resources have been utilised by the cells, the cultures die unless supplied with new medium. In practice, this is done by subculturing, i. e. transferring a small volume of existing culture to a large volume of fresh culture medium at regular intervals. In **continuous cultures**, resources are potentially infinite: cultures are maintained at a chosen point on the growth curve by the regulated addition of fresh culture medium. In practice, a volume of fresh culture medium is added automatically at a rate proportional to the growth rate of the alga, while an equal volume of culture is removed. The management of various parameters influencing the algal growth needs to be monitored as discussed below:

Temperature

The temperature at which cultures are maintained should ideally be as close as possible to the temperature at which the organisms were collected; polar organisms (<10 °C); temperate (10 – 25 °C); tropical (>20 °C). An intermediate value of 18 – 20 °C is most often employed. Temperature controlled incubators usually use constant temperature (transfers to different temperatures should be conducted in steps of 2 °C/ week), although some models permit temperature cycling. In temperate regions ambient room temperature is generally acceptable for culturing purposes.

Light

Natural light is usually sufficient to maintain cultures in the laboratory. Cultures should never be exposed to direct sunlight (which may cause phytopigment damage), and should therefore be placed next to a north-facing window (in the northern hemisphere). Artificial lighting by fluorescent bulbs is often employed for culture maintenance and experimental purposes. Light intensity should range between 0.2 – 50 % of full daylight (1660 µE/ s/ m^2), with 5 – 10 % (c. 80 – 160 µE/ s/ m^2) most often employed. Light quality (spectrum) depends on type of bulb used (see manufacturers technical data), the most common types being "cool white" or "daylight" bulbs. Light intensity and quality can be manipulated with filters. Many microalgal species do not grow well under constant illumination, and hence a light/dark (LD) cycle is used (maximum 16 : 8 LD, usually 14 : 10 or 12 : 12).

Mixing

Mixing of microalgal cultures may be necessary under certain circumstances: when cells must be kept in suspension in order to grow (particularly important for heterotrophic dinoflagellates); in concentrated cultures to prevent nutrient limitation effects due to stacking of cells and to increase gas diffusion. It should be noted that in the ocean, cells seldom experience turbulence, and hence mixing should be gentle. The following methods may be used: bubbling with air (may damage cells); plankton wheel or roller table (about 1 rpm); gentle manual swirling. Most cultures do well without mixing, particularly when not too concentrated, but when possible gentle manual swirling (once each day) is recommended.

(v) Harvesting of phytoplankton cells

In order to provide large quantities of algal foods for molluscs (like oyster, mussel etc.) and fish larvae, phytoplankton culture can also be maintained in **continuous culture system**. It consists of a 15 litre Plexiglass container with circulating water jackets for temperature control. The culture of this system is agitated by a magnetic stirring bar. The continuous flow system is controlled with micropumps which provides fresh media and a siphon gravity system for outflow. Phytoplankton cells are harvested periodically from this system depending on the phase of growth.

When the phytoplankton are cultured under controlled conditions in the laboratory, a pattern is observed during their culture period, which can be divided into several phases like *lag phase* (characterized by the absence of increment of cell number), *exponential phase* (growth phase), *stationary phase* (characterized by the constancy in cell numbers) and *death phase* (decaying phase). Harvesting of phytoplankton is usually done during the exponential phase and then they are finally transferred to carbuoys, cylinders or outdoor systems for the mass production of phytoplankton.

6.1.1. Culture of diatoms

Diatoms are important members of phytoplankton community. These are sometimes called golden algae, because their characteristic yellow-brown pigment masks their green chlorophyll. Some diatoms have radial symmetry, they are round and shaped like pill boxes, while others have bilateral symmetry and are elongate. The round diatoms float better than the elongate forms, therefore the elongate diatoms are often found in the shallow sea floor or attached to floating objects, while the round diatoms are more truly planktonic. All are found in areas of cold, nutrient rich water.

A hard, rigid transparent cell wall impregnated with silica surrounds each diatom. Pores connect the living portion of the cell inside to its outside environment. The buoyancy of the diatoms is increased by the low density of the interior of the cells and the production of oil as a storage product. Their small volume and accompanying relatively large surface area help these cells to stay afloat. Their large surface area also provides them with greater exposure to sunlight and water containing the gases and nutrients for photosynthesis and growth. In addition, some diatoms have spines or other projections that increase their ability to float.

Diatoms reproduce very rapidly by cell division; when their populations discolour the water, it is known as a bloom. Culture of these groups of phytoplankton require specialized media like TMRL medium, PM solution, Takeda medium etc.

(a) TMRL Medium

This medium which is chiefly used for the mass culture of diatoms such as *Skeletonema*, *Chaetoceros* and *Coscinodiscus* is prepared by mixing the following salts.

Potassium nitrate	1.0 gm
Sodium phosphate	0.1 gm
Ferric chloride	0.03 gm
Sodium silicate	0.02 gm

Dissolved in 10 litre of filtered seawater.

(b) PM Solution

This medium which is also normally used for the mass production of diatoms has the following composition :

Sodium nitrate	1.0 gm
Potassium phosphate	0.1 gm
Ferric chloride	0.02 gm
Sodium silicate	0.02 gm
Agrimin	0.01 gm
EDTA	0.02 gm
Thiamine (Vitamin B_1)	0.5 mg
Coblamine (Vitamin B_{12})	0.5 mg

Dissolved in 1 litre of filtered seawater.

(c) Takeda medium

For small or large scale production of diatoms such as *Biddulphia sinensis, Nitzschia longissima* and *Navicula* spp., the Takeda medium has been found to be very suitable.The composition of the medium is given below :

Salt	Amount in 1 litre	Volume added to 1 litre of filtered seawater
$NaNO_3$	200 gm	1 ml
$Na_2HPO_4.12H_2O$	50 gm	1 ml
Na_2EDTA	3 gm	1 ml
$FeCl_3.6H_2O$	0.24 gm	1 ml
$ZnCl_2$	0.03 gm	1 ml
$MnCl_2.4H_2O$	0.27 gm	1 ml
$CoCl_2.6H_2O$	0.8 gm	1 ml
$CuSO_4.5H_2O$	0.4 gm	1 ml
H_3BO_3	3.44 gm	1 ml
Tris (pH 7.4-7.5)	50 gm	1-2 ml

Medium for mass culture

For the mass culture of diatoms in open circular or concrete systems, the following medium is generally preferred :

Ammonium sulphate	2.0 gm
Urea	150 mg
Ferric chloride	200 mg
Sodium silicate	200 mg
Potassium phosphate	500 mg

Dissolved in 100 litre of filtered seawater.

6.1.2. Culture of dinoflagellates

Dinoflagellates are red to green in colour and can exist at lower light levels than diatoms, because they can both photosynthesize like a plant and ingest organic material like an animal. Their external walls do not contain silica; some are smooth and flexible but others are armoured with plates of cellulose. Dinoflagellates usually have 2 whiplike appendages or flagella that beat within grooves in the cell wall, giving the cells limited motility. Under favourable conditions, they multiply even more rapidly than diatoms to form blooms, but they are not as important as the diatoms as a primary ocean food source. Some dinoflagellates are called fire algae, because they glow with bioluminescence at night.

Dinoflagellates are hard to grow. Certain species such as *Prorocentrum micans* grow better in medium FE, which is prepared by mixing equal volmes of Guillard's medium F with Foyn's Erdschreiber medium E (Subba Rao, 1980) and autoclaved at 125 °C (2 atm pressure) for 20 min. Some dinoflagellates such as *Ceratium* grow well in Chan's medium (1978) or MET 44 (Schone and Schone, 1982) and the toxigenic *Protogonyaulax tamarensis* or *Pyrophacus steinii* in T1 medium (Ogata *et al.*, 1987; Pholpunthin *et al.*, 1999). Oceanic ultraplankton are grown successfully in medium K (Keller *et al.*, 1987).

Conway or Walne's medium is largely used for culturing phytoflagellates and has three stock solutions, viz. A, B and C whose compositions are given below:

Solution A

Sodium nitrate or potassium nitrate	10 gm
Sodium phosphate	2 gm
Ferric chloride	0.13 gm
Manganese chloride	0.04 gm
Boric acid	3.3 gm
EDTA	4.5 gm

Dissolved in 100 ml of distilled water

Solution B

Zinc chloride	2.1 gm
Calcium chloride	2.0 gm
Ammonium molybdate	2.0 gm
Copper sulphate	2.0 gm

Dissolved in 100 ml of distilled water

Solution C

Cobalamine (Vitamin B_{12})	5.0 mg
Thiamine (Vitamin B_1)	100 mg

Dissolved in 100 ml of distilled water

The medium is prepared by adding 1 ml each of the above solutions to 1 litre of filtered seawater.

6.2. Zooplankton culture

Culture of zooplankton has now become almost indispensible to run an aquacultural farm. Live fish food organisms (Fig. 6.2.1) such as brine shrimp, tubifecids, blood worms, rotifers and daphnians play a vital role in the feeding of cultivable species of fish.

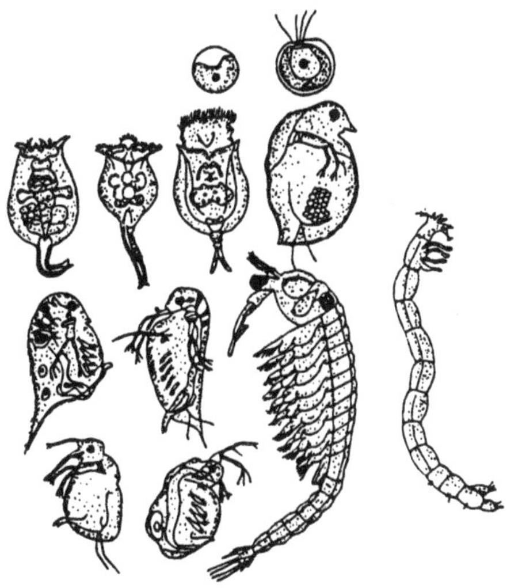

Fig. 6.2.1

Brine shrimp, popularly known as sea monkey is a common zooplankton cultured to feed tiger prawn (*Penaeus monodon*). The dormant and biconcave seeds of *Artemia* can be stored for any number of years under anaerobic conditions. However, when cysts are transferred to filtered seawater under moderate aeration and light intensity, these biconcave cysts become spherical and burst liberating the nauplii in a period of about 24 hours (Fig. 6.2.2).

These nauplii serve as instant protein-rich live food for hatchery systems for the development of invertebrate and fish larvae.

For culturing *Artemia*, ponds having depth of 40 – 70 cm are dug. This depth is normally maintained throughout the culture operation. These ponds need fertilization with chemicals and organic fertilizers in order to develop the phytoplankton and zooplankton. Among the chemical fertilizers, di-ammonium phosphate and urea are known to show better results and are applied at 50 Kg and 40 Kg per hectare respectively. For organic fertilization, cow dung or dry chicken manure is applied at 2000 Kg/ha. While the chemical fertilizers are applied weekly, the organic fertilizers are applied biweekly. Apart from fertilizers, liming is also done if the pond soil shows acidic nature.

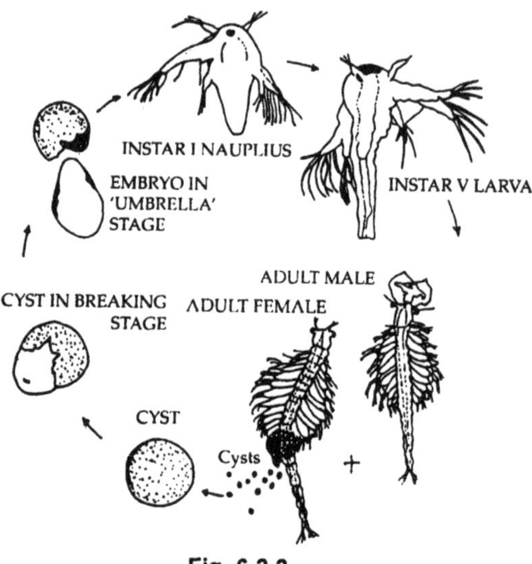

Fig. 6.2.2

In the initial stage of culture, freshly hatched nauplii from cysts are used for inoculation of the fertilized salt ponds. Depending on the water quality, the stocking density of nauplii may vary from 1 – 10 / lit. These nauplii under favourable salinity (100 –150 $^0/_{00}$) and food availability would grow into adults in two weeks. Though these adults proliferate mainly through ovo-viviparity, oviparity (cyst formation) may also be induced in these animals if and when they are put under stress by exposing them to higher salinity.

Suggested references

1. Chan, A. T. (1978). Comparative physiological study of marine diatoms and dinoflagellates in relation to irradiance and cell size. 1. Growth under continuous light. J. Phycol., 14, pp. 396-402.

2. Keller, M. D.; R. C. Selvin,; W. Claus and R. R. L. Guillard. (1987). Media for the oceanic ultraphytoplankton. J. Phycol., 23, pp. 633-638.

3. Ogata, T.T. Ishimaru and M. Kodama. (1987). Effect of water temperature and light on growth rate and toxicity change in *Protogonyaulux tamarensis*. Mar. Biol. 95, pp. 217-220.

4. Pholpunthin, P.; Y. Fukuyo, K. Matsuoka and Y. Nimura. (1999). Life history of a marine dinoflagellate Pyrophacus steinii (Schiller) Wall et Dale. Botanica Marina. 42, pp. 187-197.

5. Schone, H.K. and A. Schone. (1982). MET 44. A weakly enriched sea-water medium for ecological studies on marine plankton algae, and some examples of its application. Botanica Marina. 25, pp. 117-122.

6. Subba Rao, D.V. (1980). Measurement of primary production in phytoplankton groups by size-fractionation and germanic acid inhibition techniques. Oceanol. Acta, 3, pp. 31-42.

Internet references

- *http://www.ume.maine.edu/ssteward/phytolinks.htm*
- *http://www.seaweed.ucg.ie/cultures/CultureCollections.html*
- *http://www.reefcentral.com/vbulletin/archive/1/2001/12/1/48387*
- *http://www.sb_roscoff.fr/Phyto/collect.html*

Index